Monographs in Geometry and Topology

Volume I — *L^2 Moduli Spaces on 4-Manifolds with Cylindrical Ends*

Volume II — *The L^2 Moduli Space and a Vanishing Theorem for Donaldson Polynomial Invariants*

International Press

International Press Publications

Series in Mathematical Physics

Essays on Mirror Manifolds, Edited by S.T. Yau
Quantum Groups, S. Sternberg
75 Years of Radon Transform, Edited by S. Gindikin and P. Michor
Perspectives in Mathematical Physics, Edited by R. Penner and S.T. Yau

Series in Geometry and Topology

L^2 Moduli Spaces on 4-Manifold with Cylindrical Ends, C. Taubes
The L^2 Moduli Space and a Vanishing Theorem for Donaldson Polynomial Invariants, J. Morgan, T. Mrowka, D. Ruberman
Lectures on Harmonic Maps, R. Schoen and S.T. Yau
Lectures on Differential Geometry, R. Schoen and S.T Yau
Lectures on Low-Dimensional Topology, Edited by K. Johannson
Chern, A Great Geometer, Edited by S.-T. Yau

Series in Algebraic Geometry

Proceedings of the International Symposium on Algebraic Geometry and Related Topics, Edited by Y. Namikawa, K. Ueno, and J.-H. Yang

Series in Analysis

Integrals of Cauchy Type on the Ball, S. Gong
Proceedings of the Conference on Complex Analysis at the Nankai Institue, Edited by L. Yang

Series in Applied Physics

Proceedings of the Second International Conference on Applied Physics, Edited by D.H. Feng and T.-Y. Zhang
Physics of the Quantum Electron Solid, Edited by S.-T. Chui

Collected and Selected Works

The Collected Works of Freeman Dyson
The Collected Works of C.B. Morrey
The Collected Works of P. Griffiths
Selected Works of V.S. Varadarajan

The L^2 - Moduli Space and a Vanishing Theorem for Donaldson Polynomial Invariants

John W. Morgan
Tomasz Mrowka
Daniel Ruberman

Authors supported in part by the National Science Foundation
The third author is supported in part by a Sloan Fellowship

International Press Incorporated, Boston
P.O. Box 2872
Cambridge, MA 02238-2872

Library of Congress Catalog Card Number: 94-077174

Morgan, John W., Mrowka, Tomasz, Ruberman, Daniel
The L^2 Moduli Space and a Vanishing Theorem for Donaldson Polynomial Invariants

ISBN 1-57146-006-3

Typeset using AMS-Tex
Printed on acid free paper, in the United States of America

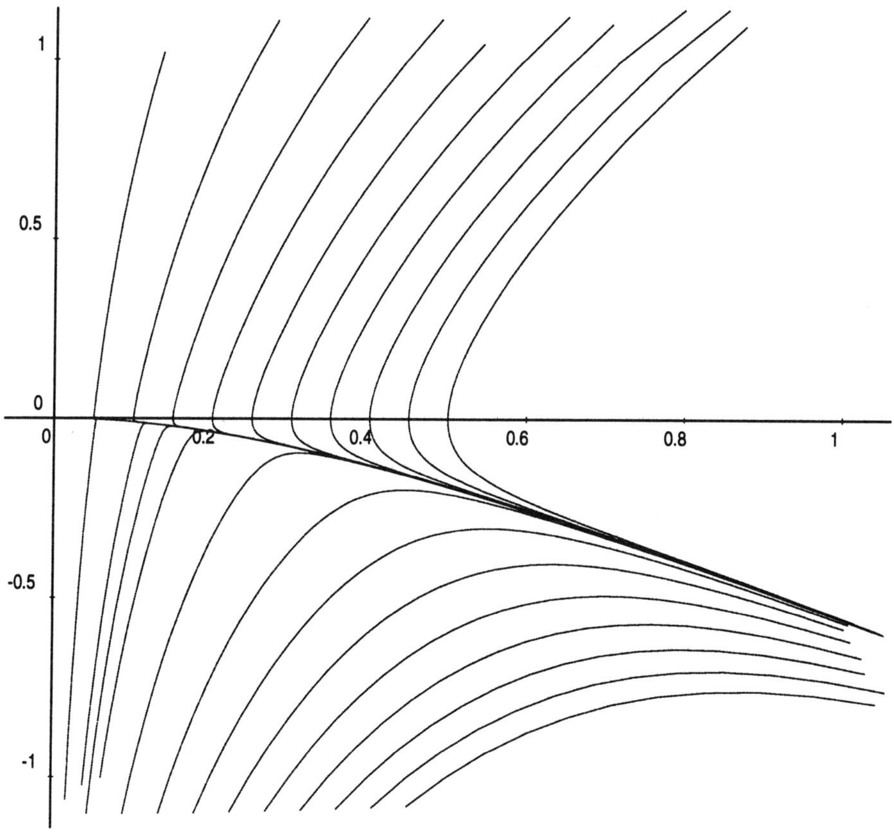

A MapleV plot of flow lines for the Chern-Simons vector field, for an Euler class 1 circle bundle over a torus. The figure shows an invariant 2-plane in $\ker d^*$; the horizontal direction lies in H^1, and the vertical direction lies in $(H^1)^\perp$. The center manifold, an invariant submanifold tangent to H^1, is clearly visible.

Contents

Preface

This book describes a setup for dealing with the general problem of trying to understand Donaldson's invariants of a smooth four manifold X by stretching its metric out along a codimension-one embedded submanifold N. The main concern here is the local description of the moduli space of finite energy anti-self-dual (ASD) connections in a principal G bundle over a four manifold with cylindrical end. There are many instances in the literature where this problem has been studied. However with the exception of [28] these studies (e.g., in the proof of the connected sum theorem and in the definition of the Floer homology theory) the variety of representations of $\pi_1(N)$ into G is smooth (or smooth after perturbation). This implies that, in those cases, finite energy connections are in fact exponentially decaying. As we shall see here, finite energy is in general a weaker condition than exponential decay. To us, the finite energy condition is the "correct" condition to study as it is preserved under Uhlenbeck limits. For example, as we stretch a tube to infinite length in the 4-manifold the Uhlenbeck limit of a sequence of ASD connections is in general an L^2 ASD connection on the limiting cylindrical end manifold. These L^2-connections will automatically be exponentially decaying only when their "boundary value" is a smooth point in the space of flat connections at infinity. This explains, from our point of view, why it suffices in previous instances cases to deal with connections with an exponential decay condition. The reader may well ask why we do not perturb the equations to achieve Floer's transversality condition. The main reason is that in important examples one already has a good understanding of the space of flat connections on the splitting three manifold (we have in mind the case of circle bundles over Riemann surfaces) and it is not clear what happens after perturbation. The analysis here is also of independent interest as it gives a rather precise method of describing the local properties of a space of finite energy solutions to a nonlinear partial differential equation on a noncompact space. This theory has been used already in two instances to prove important results in four manifold topology [15] and [19].

This book has three parts, each with several chapters. The first part deals with the L^2-moduli on a manifold with cylindrical end. The second introduces a local thickening of the L^2-moduli space and proves various local structure theorems for it. The third discusses the example of a four-manifold with an end modeled on a circle bundle over a riemann surface. This last chapter indicates the kind of results one can prove using these techniques by proving a vanishing theorem for the Donaldson invariants which contain an embedded non trivial circle bundle over a torus. A more detailed overview of the book follows.

Thoroughout this book X will denote an oriented connected riemannian four-manifold with an end isometric to a cylinder $[0,\infty)\times N$ for some oriented riemannian three-manifold N, which will usually be assumed to be connected. Part I deals with the definition of the anti-self-dual L^2-moduli space $\mathcal{M}_w(X,g)$ and the study of the basic properties of finite energy connections over $[0,\infty)\times N$. The first main technical results of this part (Theorem 4.3.1 and Corollary 4.3.3) make use of work of L. Simon [25] to find a standard form for connections on the cylinder with sufficiently small energy. Standard form means that the connections are given by a flow line for the gradient of a perturbation of the Chern-Simons function restricted to a coordinate patch in a slice for the gauge group action on the space of connections on the bundle η over N. The main content of this result is that the length of this path is bounded in terms of the energy of the connection. In particular, the path has a well-defined limit point. This leads to the existence of a well-defined boundary at infinity:

$$\partial \colon \mathcal{M}_w(X, g) \to \chi(N)$$

and the based version

$$\partial^0 : \mathcal{M}^0_w(X, g) \to \mathcal{R}(N)$$

(see Theorems 4.0.1 and 4.6.1).

The second main technical result of Part I involves a refinement of these boundary maps. We show that inside the space of connections on η for any flat connection Γ there is a center manifold $\mathcal{H}_\Gamma$ based at Γ. This manifold is tangent to $H^1(N; ad\Gamma)$ embedded inside the 1-forms by the harmonic representatives. It contains all the flat connections sufficiently close to Γ in the slice to the gauge group action at Γ and it is invariant under the unperturbed Chern-Simons flow. The technical result says that every finite energy flow line in the slice which starts sufficiently close to Γ and has sufficiently small energy in fact converges exponentially fast (with a predetermined exponent depending on Γ) to a flow line in the center manifold. In this way, we define a map from ASD connections in standard form to the stable set Y_Γ in the center manifold $\mathcal{H}_\Gamma$ which refines the boundary map at infinity given above. (See Theorem 5.2.2 and Corollary 5.2.3.)

The last chapter of Part I (Chapter 6) is of a different nature. It concerns geometric limits as we pull a 4-manifold part by introducing longer and long tubes across a given 3-manifold. As a corollary of Uhlenbeck's weak compactness theorem we show that geometric limits always exist and that these limits are L^2-connections (see Theorem 6.1.1). We then refine the the notion of geometric limits so that one can always arrive at a situation where the geometric limit has the same energy as the connections in the sequence, see Theorem 6.3.3. The main result of this chapter is that in this case the geometric limit has what we call compatible boundary values: this simply means that the flat connections along components of infinity that are to be glued up to form the original 4-manifold are in fact isomorphic flat connections, (see Proposition 6.3.1).

In Part II we study the local nature of $\mathcal{M}_{\ell,w}(X,g)$ in more detail. To do this it is useful to embed the L^2-moduli space in a manifold analogous to the way the stable set of the critical set sits in the center manifold. To do this, we define *modified ASD equations* whose moduli space, called the *thickened moduli space*, projects to a neighborhood of the origin in the center manifold. The thickened moduli space seems to provide the correct geometric context for a general gluing theorem for ASD connections, although we do not treat this topic in this book. The main result of this part is a generic metrics theorem for the thickened moduli space (cf Theorem 9.0.1). For a generic metric this thickened moduli space is a smooth manifold mapping generically to this neighborhood with fibers which are generically manifolds whose dimension is the formal dimension calculated using the technique of Atiyah-Patodi-Singer. Furthermore, the L^2-moduli space sits in the thickened moduli space as the preimage of the stable set in the center manifold under the projection. In the special case when the stable set is a stratified subset of the center manifold, Sard's theorem for a generic metric implies that the submersion will be transverse to the stable set. In this case we will produce a stratified structure locally on the L^2-moduli space. In general, we do not know whether the stable set is always stratified, and if so whether this stratification is independent of the choice of center manifold, and if it varies nicely as we move the central flat connection. If that is the case, then the local stratifications of the L^2-moduli space will fit together to give the L^2-moduli space a global stratified structure. One example where this good case occurs is when the riemannian 3-manifold is an S^1-bundle over a surface with a particular class of riemannian metrics.

The third part shows how the techniques developed in the first two parts can be used to study the case of splitting a closed four-manifold along the total space of a nontrivial circle bundle N over a riemann surface. First we review the Kuranishi deformation picture for flat connection Γ on a bundle η over a three-manifold and find that in this case the deformation theory gives us a vector field Z in a neighborhood of the origin in $H^1(N; Ad\Gamma)$. This vector field is easier to compute than the restriction of ∇f_Γ to the center manifold. Some general results are given comparing these two vector fields. Then we turn to explicit computations in the case of a nontrivial circle bundle N over a riemann surface. In this case one can find a nonpositive function on $H^1(N; Ad\Gamma)$ whose gradient is a sufficiently good approximation to ∇f_Γ. We can show that for one orientation of N the stable set of the critical set contains a neighborhood of Γ, while for the other orientation the stable set is the same as the critical set. We then use the general machinery of the previous chapters to prove the local structure theorems mentioned above for the L^2-moduli space. One finds that for a generic metric the L^2-moduli space is a smooth manifold if the Euler class of the circle bundle is positive. If the Euler class is negative, then the L^2-moduli space maps to the character variety in such a way that the preimage of each stratum of the character variety is a smooth manifold. This structure will be important in our applications.

The application of these results we discuss in this book is a vanishing theorem for the Donaldson invariants. Recall that for compact oriented 4-manifold A with boundary one denotes by $b_2^+(A)$ the dimension of any maximal positive definite subspace of $H^2(A, \partial A; R)$ for the intersection pairing.

For a closed 4-manifold M we denote by $d(k) = d_M(k)$ the number

$$4k - \frac{3}{4}(\chi(M) + \sigma(M)).$$

It is one-half of the formal dimension of the moduli space of ASD connections on a principal $SU(2)$-bundle over M with Chern class $c_2 = k$.

Here is our main result, to be established in Chapter 16.

THEOREM. *Let M be a closed oriented connected 4-manifold and suppose that we have a decomposition $M = P \cup_{\overline{N}} Q$ where $\overline{N}$ is a circle bundle over a torus.*

Suppose that the Euler class of $\overline{N}$ *is even (and nonzero) and that* $b_2^+(P)$ *and* $b_2^+(Q)$ *are greater than* 0*. Then for all* $k > 0$ *the Donaldson polynomial invariant* $\gamma_k(M)$ *associated to* $SU(2)$*-bundles with* $c_2 = k$ *vanishes when evaluated on* $d(k)$ *classes in the image of* $H_2(M - \overline{N}; Q) \to H_2(M; Q)$.

As an easy corollary of this result we have:

COROLLARY. *Suppose that* M *is a closed, oriented* 4*-manifold with* $b_2^+(M) > 1$ *and* $H_1(M) = 0$*. Suppose that* $T^2 \subset M$ *is a smoothly embedded torus of self-intersection at least* 2*. Then the Donaldson polynomial invariants* $\gamma_k(M)$ *vanish.*

Throughout this paper we use the following notation: G is a compact simple Lie group, $E \to X$ is a principal G-bundle; $\eta \to N$ is the restriction of E; $\mathcal{A}_k(\eta)$ is the space of L_k^2-connections on η; $\mathcal{G}_{k+1}(\eta)$ is the group of L_{k+1}^2-changes of gauge; $\mathcal{B}_k(N)$ is the quotient of $\mathcal{A}_k(\eta)$ be $\mathcal{G}_{k+1}(\eta)$. We denote by $\Omega^i(N)$ the C^∞-forms with values in $ad\eta$ and by $\Omega_k^i(N)$ the L_k^2-forms with values in $ad\eta$. The boundary of an oriented manifold X will be oriented so that the outward normal followed by the orientation of ∂X is the orientation of X.

Acknowledgment We began working on this book at a time when, in one capacity or another, we were all at Harvard, and we would like to thank the Mathematics Department for its support and hospitality. After we realized that it was necessary to consider the L^2-moduli space in order to study general splittings, we had a long series of discussions with Cliff Taubes concering many of the issues covered in this paper. Out of these discussions arose Taubes' paper [28] and this paper. We wish to thank Cliff for generously sharing his time and insights with us. We also thank Ron Fintushel and Ron Stern for their critical reading and useful comments, as well as A. Stipicz and S.-M. Wu for help with the proofreading.

Part I

The L^2-moduli space for manifolds with cylindrical ends

CHAPTER 1

Introduction: The L^2-moduli space

This part concerns the L^2-moduli space, that is the moduli of finite energy ASD connections on a principal bundle E with structure group a given compact simple Lie group G over a complete riemannian 4-manifold with a cylindrical end. There are many instances in the literature where ASD connections on such bundles have been studied. However with the exception of [28] these studies (e.g., in the proof of the connected sum theorem and in the definition of the Floer homology theory) the variety of representations of $\pi_1(N)$ into G is smooth. This implies that in those cases finite energy connections are in fact exponentially decaying. As we shall see here, finite energy is in general a weaker condition. To us, the finite energy condition is the "correct" condition to study as it is preserved under Uhlenbeck limits. For example as we stretch a tube to infinite length in the 4-manifold the Uhlenbeck limit of a sequence of ASD connections is in general an L^2 ASD connection on the limiting cylindrical end manifold. This explains, from our point of view, why it suffices in previous instances to deal with connections with an exponential decay condition.

Once we drop the exponential decay condition, the entire theory has to be recast. Fix a closed oriented, riemannian 3-manifold (N, g_N) and a complete, oriented riemannian 4-manifold (X, g_0) with a cylindrical end which is orientation-preserving isometric to $[0, \infty) \times N$. We call g_0 the standard metric. Let E be a principal G-bundle over X. Assuming as we shall that no component of X is closed, such bundles are determined up to isomorphism by a characteristic class $w = w(E) \in H^2(X; \pi_1(G))$. For any complete riemannian metric g on X we define the *energy* of a connection ω to be

$$- \int_X trF_\omega \wedge *F_\omega = \int_X |F_\omega|^2 = ||F_\omega||^2_{L^2}$$

where the trace is the trace of the adjoint action of the Lie algebra of G on itself. Of course, this energy may be infinite. We define the L^2-moduli space $\mathcal{M}_w(E, g)$ to be the space of gauge equivalence classes of $L^2_{2,loc}$ g-ASD connections on E with the property that energy is finite. The gauge transformations are $L^2_{3,loc}$ bundle automorphisms of E. The topology on the space of $L^2_{2,loc}$ g-ASD connections is the weakest topology containing the $L^2_{2,loc}$ topology in which the energy function is continuous. There is an analogous notion of the based L^2-moduli space, where we take as base point a point of the form $(0, n) \in [0, \infty) \times N \subset X$.

It turns out that the energy for such connections takes on a discrete set of values, see Corollary 4.1.3. Thus, we can consider the subspace $\mathcal{M}_{\ell,w}(X, g)$ of ASD connections of fixed energy $2\pi^2\ell$. Notice that $\ell \geq 0$.

For the purposes of this introduction we shall pretend that all the the flat connections on $\eta = E|N$ are irreducible, and we shall work in the unbased context. In the detailed analysis, to handle the case of reducible flat connections on η, we work equivariantly with the space of based connections. This introduces no new conceptual issues.

Our goal is to give a local description of the L^2-moduli space $\mathcal{M}_{\ell,w}(X, g)$ of ASD connections of fixed energy $2\pi^2\ell$. Notice that $\ell \geq 0$.

For the purposes of this introduction we shall pretend that all the the flat connections on $\eta = E|N$ are irreducible, and we shall work in the unbased context. In the detailed analysis, to handle the case of reducible flat connections on η, we work equivariantly with the space of based connections. This introduces no new conceptual issues.

Our goal is to give a local description of the L^2-moduli space $\mathcal{M}_{\ell,w}(X,g)$ for all metrics sufficiently close to g_0. From the definition, little is obvious *a priori*. The key to our local description is the relationship between the ASD equations in the cylinder and the gradient flow of the Chern-Simons function on the space of connections on the 3-manifold at infinity. Let us discuss this relationship. Fix a flat connection Γ on η, and a slice S_Γ for the action of the gauge group on the space of connections on η. Each connection ω on E determines a family of gauge equivalent paths of connections on η, via a bundle isomorphism $[0,\infty)\times\eta \xrightarrow{\cong} E|([0,\infty)\times N)$. Choosing such a bundle isomorphism, the connection ω determines a path $A_\omega(t)$ of connections on η. A different bundle isomorphism will give a different path of connections on η, one which differs from $A_\omega(t)$ by a one-parameter family of gauge transformations of η. Let us suppose that we have a bundle isomorphism such that this path $A_\omega(t)$, or some part of it, is a path in S_Γ. If ω is g_0-ASD then $A_\omega(t)$ is an integral curve for the gradient flow of the restriction f_Γ of the Chern-Simons function to the slice S_Γ. More generally, if the metric g is a perturbation of the product metric g_0 then $A_\omega(t)$ is an integral curve of a (time-dependent) perturbation of the gradient flow for f_Γ (see Theorem 2.6.3).

The first consequence of this is that for metrics g sufficiently close to g_0 any finite energy g-ASD connection ω has a well-defined asymptotic limit, see Theorem 4.0.1. More precisely, given such an ω we can find a flat connection Γ and $T_0\geq 0$ and a bundle isomorphism $[0,\infty)\times\eta \xrightarrow{\cong} E|([0,\infty)\times N)$ such that the path $A_\omega(t), T_0\leq t<\infty$, induced from ω' is a path in S_Γ. This path converges to a flat connection Γ' in S_Γ. Of course, by Uhlenbeck's weak compactness result since ω has finite energy the path $A_\omega(t)$ limits to the subset of flat connections on η. But there is no naïve reason that there should be a well-defined limiting flat connection up to change of gauge. Conceivably the path $A_\omega(t)$ could spiral into the space of flat connections without a limit. That this does not happen is our first main result about $\mathcal{M}_{\ell,w}(X,g)$. Not only do we deduce the existence of a limiting connection, but we also prove strong limiting results. These limiting results imply, for instance, that the length of the path $A_\omega(t)$ is finite, and bounded by a power of the energy of the path, see Theorem 4.3.1. All of this is achieved by using results of L. Simon's concerning asymptotics of finite energy integral curves for gradient flow equations associated to real analytic functionals with elliptic linearization and hence is based on the fundamental work of Lojasiewicz on real analytic functions. Because of these results, for any metric g sufficiently close to g_0, we can define a continuous map ∂_∞ from $\mathcal{M}_{\ell,w}(X,g)$ to the character variety $\chi(N)$ of $\pi_1(N)$, see Theorem 4.0.1. This map associates to each finite energy g-ASD connection its asymptotic flat connection.

This approach also yields a locally defined refinement of the map ∂_∞. Suppose that ω is a finite energy ASD connection on E of energy ℓ whose asymptotic flat connection is Γ. Then there is a neighborhood $U(\omega)$ of $[\omega]$ in $\mathcal{M}_{\ell,w}(X,g)$ and $T_0 \geq 0$ such that if $[\omega'] \in U(\omega)$ then, there is a bundle isomorphism $[0,\infty) \times \eta \xrightarrow{\cong} E|([0,\infty) \times N)$ such that the path $A_{\omega'}(t)$, $T_0 \leq t < \infty$, induced by ω' lies in the slice S_Γ. Furthermore, this path is of finite length and limits to a flat connection, which is the asymptotic value of $[\omega']$. As we noted above, this path in S_Γ is an integral curve for a (time-dependent) perturbation of ∇f_Γ. Assuming, as we shall from now on, that the metric g approaches the standard metric g_0 exponentially fast, the perturbation term goes to zero exponentially fast, with the same exponent. The local refinement of ∂_∞ mentioned above is a consequence of the center manifold theorem in this context. First notice that the set of flat connections in S_Γ is exactly the critical set for the ∇f_Γ. Secondly, notice that the tangent space to S_Γ at $\{\Gamma\}$ is $Ker\, d_\Gamma^* \subset \Omega^1(N)$, and that the Hessian of f_Γ at Γ is the essentially self-adjoint unbounded operator $- * d_\Gamma | Ker d_\Gamma^*$. The center manifold theorem (cf. Theorem 5.1.1) says that there is a finite dimensional submanifold $\mathcal{H}_\Gamma$, tangent at $\{0\}$ to the null-space of $- * d_\Gamma | Ker d_\Gamma^*$. This manifold is invariant under the flow generated by ∇f_Γ, and it has the following property. Suppose that

$[\omega'] \in U(\omega)$ and we have chosen the bundle isomorphism as above so that the path $A_{\omega'}(t)$ lies in S_Γ. Then the path $A_{\omega'}(t)$ approaches exponentially fast to an integral curve $B_{\omega'}(t)$, $T_0 \leq t < \infty$, for the restriction of ∇f_Γ to $\mathcal{H}_\Gamma$. The exponent in this approach is $e^{-\mu_\Gamma^-(t-T_0)}$ where μ_Γ^- is the absolute value of the first negative eigenvalue for $- * d_\Gamma|\mathcal{K}_\Gamma$. (All this is assuming, as we shall, that the exponent in the exponential approach of g to g_0 is larger than μ_Γ^-.) In particular, $B_{\omega'}(t)$ is an integral curve for ∇f_Γ which also has finite length and which limits to the same flat connection as $A_{\omega'}(t)$. Fixing T_0 sufficiently large, and associating to $[\omega']$ the point $B_{\omega'}(T_0) \in \mathcal{H}_\Gamma$ defines a continuous map $\overline{\Upsilon}$ from $U(\omega) \subset \mathcal{M}_{\ell,w}(X,g)$ to the stable set $Y_\Gamma \subset \mathcal{H}_\Gamma$ of the critical set $C_\Gamma \subset \mathcal{H}_\Gamma$ of the gradient flow for f_Γ restricted to $\mathcal{H}_\Gamma$. This map refines the map ∂_∞ because the limiting value of the flow line through $\overline{\Upsilon}([\omega'])$ is $\partial_\infty([\omega'])$.

Of course, each flow line $B_{\omega'}(t)$ determines a finite energy g_0-ASD connection on $[T_0, \infty) \times \eta$. Thus, the fibers of $\overline{\Upsilon}$ are spaces of finite energy connections which differ by an error term of size $e^{-\mu_\Gamma^-(t-T_0)}$ from a fixed finite energy g_0-ASD connection in the cylinder. In the special case when the corresponding flow line on the center manifold is constant (necessarily at a flat connection) the fiber is the moduli space of connections exponentially decaying to the given flat connection. If Γ is a smooth point of the representation variety then the center manifold consists entirely of flat connections, i.e., $C_\Gamma = W_\Gamma^s = \mathcal{H}_\Gamma$, the restriction of the gradient flow to the center manifold is trivial, and all fibers are moduli spaces of connections exponentially decaying to flat connections. The new phenomena occur near the singular points of the representation variety. If $[\Gamma]$ is a singular point of the representation variety, then it can happen that the stable manifold W_Γ^s is larger than the critical set. As a result there will be ASD connections with asymptotic value Γ which do not have exponential decay to Γ. The exact nature of the stable manifold W_Γ^s and the gradient flow on it are important aspects, in fact dominating aspects, of the local structure of $\mathcal{M}_{\ell,w}(X,g)$ near $\partial_\infty^{-1}([\Gamma])$.

The fibers of $\overline{\Upsilon}: U(\omega) \to W^s_\Gamma$ can be studied by the techniques previously used to study the moduli spaces of exponentially decaying connections. In particular, one can compute the formal dimension of these fibers as the index of a certain elliptic operator on a weighted Sobolev space, see Corollary 8.3.2. The computation of this index uses familiar techniques from Atiyah-Patodi-Singer, see Proposition 8.4.1.

The last section of this part deals with Uhlenbeck limits of ASD connections on closed riemannian four manifold as a tube in X is stretched to infinite length. As we mentioned above, given a sequence of metrics with tube length going to infinity , a sequence of connections ASD with respect to this sequence of metrics has an Uhlenbeck limit which is an ASD connection on the associated cylindrical end manifold. If there is no energy lost in the tube in this limit then the boundary values on the ends of the cylindrical end manifold agree (cf. Theorem 6.3.1).

CHAPTER 2

The Chern-Simons flow and ASD connections on a cylinder

As we indicated in the introduction, much of the analysis of the L^2-moduli space revolves around the relationship between the gradient flow of the Chern-Simons function on the space $\mathcal{A}(\eta)$ and the ASD equation for connections on $[0,\infty)\times\eta$. This relationship is given in in Equation 4 in the case of the product metric $g_0 = g_N + dt^2$ and is given in Equation 5 in the case of a general metric. In practice we work not with the space $\mathcal{A}(\eta)$ of connections on η but rather with the slices in this space for the action of the group of gauge transformations. Since we shall always be working in a neighborhood of the subspace of flat connections, it will suffice for our purposes to consider only slices centered at flat connections. Let Γ be a C^∞ flat connection, and let U_Γ be a slice for the action of the gauge group on $\mathcal{A}(\eta)$ through Γ. We show how a connection on $[0,\infty)\times\eta$ with the property that its restriction to each $\{t\}\times\eta$ is gauge equivalent to a connection in U_Γ can be written on the entire cylinder, up to gauge equivalence, in terms of a path in U_Γ, see Lemmas 2.4.3 and 3.1.3. Next, we turn our attention to the restriction f_Γ of the Chern-Simons function to U_Γ, and give a formula for the gradient of this function. All of this material on slices culminates in the last section of this chapter, where we study the flow equation in U_Γ determined by the ASD equation for connections on $[0,\infty)\times\eta$. In the case of the product metric on the cylinder, the resulting equation is simply the gradient flow equation for f_Γ. When the metric is a perturbation of the product metric, the flow equation is more complicated, but it should be viewed as a perturbation of this gradient flow equation. Theorem 2.6.3, which is the main result of this chapter, gives the precise formulation.

The proof of Theorem 2.6.3 and various technical refinements are given in the next chapter.

2.1. The basics of the Chern-Simons function

In this section we introduce the Chern-Simons function, compute its gradient at a connection A in terms of the curvature F_A, and give a formula for its Hessian at a critical point.

Consider the four dimensional characteristic class $p_1(adE) \in H^4(X;R)$ of the adjoint bundle associated to $E \to X$. Given a connection ω on E, a closed form representative for this class is given by the Chern-Weil formula

$$p_1(ad\omega) = -\frac{1}{8\pi^2} tr(F_\omega \wedge F_\omega)$$

where we use tr to denote the trace of the adjoint representation of the Lie algebra of G on itself. [If $G = SU(2)$, then we can compare this characteristic class with c_2 of the C^2-bundle V associated to the fundamental representation and obtain that $c_2(V) = -\frac{1}{4} p_1(adE)$. We also see that the trace in the adjoint representation is four times the trace in the fundamental representation. This is all consistent since

$$c_2(E) = \frac{1}{8\pi^2} \tilde{tr} F_\omega \wedge F_\omega$$

where $\tilde{tr}$ denotes the trace in the fundamental representation. In this case $\ell = c_2$, thus explaining why we have chosen this normalization for ℓ.] Notice that if ω is ASD then we have that the energy of ω is equal to $-8\pi^2 \int_X p_1(ad\omega) = 2\pi^2 \ell$.

Let us fix a C^∞ connection Γ on η. We define the Chern-Simons function with respect to Γ, CS_Γ, as follows. Let A be a connection on η. Write $A = \Gamma + a$ for $a \in \Omega^1(N; ad\eta)$. Then

$$CS = CS_\Gamma(A) = \int_N tr\left(2a \wedge F_\Gamma + a \wedge d_\Gamma a + \frac{2}{3} a \wedge a \wedge a\right).$$

If $\varphi\colon \eta \to \eta$ is a change of gauge then

$$CS(\varphi^*(A)) - CS(A) = -8\pi^2\langle p_1(adE), [S^1 \times N]\rangle. \tag{1}$$

where E is the mapping torus of the gauge transformation ϕ.

Thus, the Chern-Simons function descends to a function

$$\overline{CS}\colon \mathcal{B}(N) \to R/Z \cong S^1.$$

It is also convenient to consider the universal covering $\tilde{\mathcal{B}}(N)$ of $\mathcal{B}(N)$ which is obtained by dividing out by the component of the identity of the gauge group. The Chern-Simons function induces a function

$$CS\colon \tilde{\mathcal{B}}(N) \to R.$$

Let Y be a compact oriented 4-manifold, let $E \to Y$ be a principal G-bundle, let ω be a connection on E, and let $\hat{\omega}$ be a connection on E whose restriction to ∂Y is Γ_0. Then the Chern-Simons Theorem [8] says that

$$\int_Y p_1(ad\omega) - p_1(ad\hat{\omega}) \equiv CS_\Gamma(\omega|\partial Y). \tag{2}$$

If ω is a connection on $[0,1] \times \eta$ whose restriction at each $\{t\} \times \eta$ is a flat connection then the integrand on the left-hand side of Equation 2 is 0. It follows then that CS is constant on components of the subspace of flat connections. From this, the compactness of $\chi(N)$, and Formula 1 it follows that the values of CS on flat connections form a discrete subset of R.

Next, we compute explicit formulas for the gradient and the Hessian of CS. These formulas are standard and are given in [27].

LEMMA 2.1.1. *The tangent space to $\mathcal{A}_1(\eta)$ at a connection A is identified with $\Omega^1_1(N; ad\eta)$.*

(1) *The differential of CS is given by*

$$D_A(CS)(\tau) = \langle - * F_A, \tau\rangle_{L^2}.$$

Using the L^2-inner product on the tangent spaces, we rewrite this equation as

$$\nabla_A(CS) = - * F_A$$

with the understanding of course that $\nabla_A(CS)$ is not a tangent vector but is only in the L^2-completion of the tangent space.

(2) *Let $\Gamma \in \mathcal{A}_1(\eta)$ be a critical point for CS. Then the Hessian of CS at Γ is given by*

$$Hess_\Gamma(CS)(\tau_1, \tau_2) = \langle \tau_1, - * d_\Gamma \tau_2 \rangle_{L^2}$$

for any pair τ_1, τ_2 of tangent vectors. The adjoint of the Hessian is the (unbounded) essentially self-adjoint operator

$$- * d_\Gamma \colon \Omega^1(N; ad\eta) \to \Omega^1(N; ad\eta).$$

The spectrum of this operator is real and discrete and unbounded in both directions. Every nonzero eigenvalue is of finite multiplicity. $d_\Gamma(\Omega^0(N; ad\eta)) \subset Ker(- * d_\Gamma)$. The L^2-orthogonal complement of $d_\Gamma(\Omega^0(N; ad\eta))$ in $Ker(- * d_\Gamma)$ is finite dimensional. The span of the eigenspaces is dense in $Ker(- * d_\Gamma)$.

DEFINITION 2.1.2. *Let μ_Γ^- (resp., μ_Γ^+) be the absolute value of the first negative (resp., positive) eigenvalue of $- * d_\Gamma$.*

2.2. The relationship with the ASD Equation on the cylinder

The main reason that the Chern-Simons function and its gradient are important in our discussion is their relation to the ASD equation for connections on a bundle over a cylinder. To facilitate the discussion we introduce the following notation. Let ω be a connection on $[T_0, T_1] \times \eta$. We can write

$$\omega = A(t) + \alpha(t)dt$$

where A is a 1-parameter family of connections on η and α is a 1-parameter family of 0-forms on N with values in $ad\eta$. Then

$$F_\omega = F_A + dt \wedge \left(\frac{\partial A}{\partial t} - d_A(\alpha) \right).$$

Thus, the g_0-ASD equation written in terms of A and α is

$$\frac{\partial A}{\partial t} = - * F_A + d_A(\alpha) \tag{3}$$

where $*$ is the Hodge $*$-operator for forms on N with respect to the fixed metric g_N. Notice that the right-hand side of this equation differs from $- * F_A$ by a vector in the tangent to the orbit of the gauge group through A. Thus, this flow equation is gauge equivalent to the gradient flow equation for CS:

$$\frac{\partial A}{\partial t} = - * F_A = \nabla_A(CS) \tag{4}$$

in the sense that if $A(t)$ satisfies Equation 3 then there is a 1-parameter family of changes of gauge $\sigma(t)$ such that the flow $B(t) = \sigma(t)^* A(t)$ satisfies Equation 4.

The discussion so far has been for the product metric $g_0 = g_N + dt^2$ on $[0, \infty) \times N$. We consider metrics of the form $g = g_0 + h$ where h is a symmetric 2-tensor on $[0, \infty) \times N$ bounded in the C^3-norm. The precise bound is determined in Theorem 2.6.3. But one condition is that h is sufficiently small so that $g_0 + h$ is a C^3-riemannian metric on $[0, \infty) \times N$. As we indicated above the ASD equation with respect to g_0 for a 2-form $B + dt \wedge \beta$ is $\beta = - * B$ where $*$ is the Hodge star operator on N with respect to g_N. The anti-self-dual condition for the metric $g_0 + h$ can be written as:

$$B + dt \wedge \beta \ is\ g_0 + h\ anti-self-dual\ if\ and\ only\ if\ \beta = \phi_h(- * B)$$

where for each $(t, n) \in [0, \infty) \times N$

$$\phi_h(t, n) \colon T^* N_{(t,n)} \to T^* N_{(t,n)}$$

is a linear automorphism which depends real analytically on the 2-tensor $h(t, n)$. For such a ϕ we define $||\phi||_{C^0}$ to be the maximum of the pointwise (finite dimensional) operator norms $||\phi(t,n)||_{C^0}$ measured with respect to g_0. Similarly for $||\phi||_{C^i}$ for $i = 1, 2, 3$.

If we write a connection $\omega = A(t) + \alpha(t)dt$ as before, then the g-ASD equation becomes

$$\frac{\partial A}{\partial t} = -\phi_h(*F_A) + d_A(\alpha). \tag{5}$$

Because ϕ_h depends real analytically on h and equals the identity at any point where $h = 0$ we have the following:

LEMMA 2.2.1. *There are constants $C, b > 0$ such that for all h with $||h||_{C^3} \leq b$ we have $||\phi_h - Id||_{C^3} \leq C||h||_{C^3}$ and $||\phi_h||_{C^3} \leq 2$.*

2.3. Slices for the gauge group action on $\mathcal{A}(\eta)$

So far we have related the ASD Equation on the cylinder $[0, \infty) \times \eta$ with the gradient flow for the Chern-Simons functional on the space of connections on η. Rather than working with the entire space of connections on η we need to work with a slice for the action of the gauge group. We find it convenient to center our slices at flat connections.

Let us recall the basic facts about slices for the gauge action on the space of connections on η. Let $\Gamma \in \mathcal{A}_{\frac{3}{2}}(\eta)$ be a connection. We set

$$\mathcal{K}_\Gamma = Kerd_\Gamma^* \cap \Omega^1_{\frac{3}{2}}(N).$$

A *slice for the action of* $\mathcal{G}_{\frac{5}{2}}(\eta)$ is given by

$$S_\Gamma = \{\Gamma\} + \mathcal{K}_\Gamma \subset \mathcal{A}(\eta).$$

The subspace S_Γ is invariant under the action of $Stab(\Gamma)$, and an open $Stab(\Gamma)$-invariant neighborhood U_Γ of $\{\Gamma\}$ in S_Γ has the property that the quotient map induces a homeomorphism $U_\Gamma/Stab(\Gamma)$ to an open neighborhood of $[\Gamma] \in \mathcal{B}_{\frac{3}{2}}(N)$.

DEFINITION 2.3.1. *Any $Stab(\Gamma)$-invariant neighborhood U_Γ as above is called* a coordinate patch centered at Γ. *We denote connections in U_Γ by A and we denote by a the one form $A - \Gamma$.*

It is important to be able to compare distances between orbits and distances in the slice.

LEMMA 2.3.2. *There is a constant $K_1 > 0$ depending on Γ so that for all sufficiently small coordinate patches U_Γ centered at Γ the following estimate holds. For all A and A' connections on η gauge equivalent to $\Gamma + a, \Gamma + a' \in U_\Gamma$ then*

$$min_{g \in Stab(\Gamma)} ||a - g^* a'||_{L^2_{\frac{3}{2}}} \leq K_1 ||A - A'||_{L^2_{\frac{3}{2}}} \tag{6}$$

We shall not prove this lemma here: essentially it boils down to showing that the embedding of the slice theorem

$$U_\Gamma \times_{Stab(\Gamma)} \mathcal{G}(\eta) \hookrightarrow \mathcal{A}_{\frac{3}{2}}(\eta)$$

has differential which is an isomorphism with norm bounded below by a positive constant.

Let us consider now other useful norms on the slice. First, let us suppose that Γ is irreducible. Then provided that U_Γ is sufficiently small, all $A \in U_\Gamma$ are also irreducible. The restriction of the L^2-inner product on the space $\Omega^1(N)$ induces an incomplete metric on $\mathcal{K}_\Gamma$ and hence on U_Γ. There is a more natural riemannian metric on U_Γ, given by the pullback to U_Γ of the L^2 metric on the quotient space $\mathcal{B}(N)$. In this metric, the inner product of two tangent vectors τ_1, τ_2 at a connection $A \in U_\Gamma$ is

$$\langle p_A(\tau_1), p_A(\tau_2) \rangle_{L^2}$$

where p_A is the L^2-orthogonal projection $\mathcal{K}_\Gamma \to \mathcal{K}_A$.

We want a similar description when Γ is reducible. We must deal with the fact that as $A \in U_\Gamma$ varies the subspaces $\mathcal{K}_A$ do not form a locally trivial family. For U_Γ sufficiently small the spaces $d_A\left(Ker(\Delta_\Gamma)^\perp\right)$ do form a locally trivial family. Let

$\mathcal{K}'_A \supset \mathcal{K}_A$ be the orthogonal complements, which also form a locally trivial family of subspaces. Of course, $\mathcal{K}'_\Gamma = \mathcal{K}_\Gamma$, and if Γ is irreducible then for each $A \in U_\Gamma$ we have $\mathcal{K}'_A = \mathcal{K}_A$.

For general Γ, a natural riemannian metric on U_Γ is given as follows. The inner product of two tangent vectors τ_1, τ_2 at a connection $A \in U_\Gamma$ is

$$\langle p_A(\tau_1), p_A(\tau_2)\rangle_{L^2}$$

where p_A is the L^2-orthogonal projection $\mathcal{K}_\Gamma \to \mathcal{K}'_A$. This riemannian metric on U_Γ is invariant under $Stab(\Gamma)$. At any irreducible connection $A \in U_\Gamma$ the projection from U_Γ to the quotient space $\mathcal{B}(N)$ is a riemannian submersion.

We denote this inner product by $\langle\cdot,\cdot\rangle_{\{\Gamma,A\}}$.

REMARK 2.3.3. *By [12] pg. 198 for any $k \geq 3/2$ if σ is an L^1 gauge transformation and if A and σ^*A are in $\mathcal{A}_k(\eta)$ then σ is in $\mathcal{G}_{k+1}(\eta)$. Hence there is a natural inclusion $\mathcal{B}_k(N) \hookrightarrow \mathcal{B}_{\frac{3}{2}}(N)$. Furthermore, if Γ an L^2_k-connection and if $A \in U_\Gamma \subset \mathcal{A}_{\frac{3}{2}}(\eta)$ projects to $[A] \in \mathcal{B}_k(N)$ then $A \in \mathcal{A}_k(\eta)$.*

2.4. Connections on $[0,\infty) \times N$ and paths in $\mathcal{B}(N)$

We find it convenient to introduce the following definition.

DEFINITION 2.4.1. *An automorphism τ of the bundle $[T_0,T_1) \times \eta$ is said to be the pullback of an automorphism τ_0 of η if it $\tau|\{t\} \times \eta = \tau_0$ for all $t \in [T_0,T_1)$.*

We need a basic result relating connections on $[T_0,T_1)\times\eta$ to paths of connections in a slice U_Γ. The following definition facilitates the discussion.

DEFINITION 2.4.2. *Let U_Γ be a coordinate patch centered at Γ and let I be a subinterval of $[0,\infty)$ An L^2_2 connection $\omega = A(t) + \alpha(t)dt$ on $I \times \eta$ is said to be in standard form with respect to U_Γ if for all $t \in I$*

(a) *$A(t) \in U_\Gamma$ and*
(b) *$\alpha(t) \in Ker(\Delta_\Gamma)^\perp$*

A connection is said to be in standard form with respect to Γ *if there is some coordinate patch* U_Γ *as above.*

LEMMA 2.4.3. *Let* Γ *be a* C^∞*-connection on* η*. Let* U_Γ *be a coordinate patch centered at* Γ*, and let* $I \subset [0,\infty)$ *be an interval. Let* ω *be a* C^∞*-connection on* $I \times \eta$*. Suppose for each* $t \in I$ *that* $\omega|(\{t\} \times \eta)$ *is* C^∞*-gauge equivalent to a connection in* U_Γ*. Then there is a* C^∞*-gauge transformation* σ *defined on* $I \times \eta$ *so that the* C^∞ *connection* $\sigma^*\omega = A(t) + \alpha(t)dt$ *is in standard form with respect to* U_Γ*.*

The gauge transformation σ *is unique up to composition with the pullback of an automorphism* $\tau_0 \in Stab(\Gamma)$*.*

REMARK 2.4.4. : *It follows from Remark* 2.3.3 *that the connection* A *obtained in the conclusion of the previous lemma defines a* C^∞ *path of* C^∞ *connections in* U_Γ*.*

First let us establish a weaker result.

LEMMA 2.4.5. *For* Γ*,* U_Γ *and* ω *as in the hypothesis of Lemma* 2.4.3 *there is a* C^∞*-gauge transformation* σ *so that* $\sigma^*\omega = A(t) + /alfa(t)dt$ *where* $A(t) \in U_\Gamma$*.*

Proof. We write

$$\omega = B(t) + \beta(t)dt$$

for $B(t)$ a C^∞-path of C^∞-connections in $\mathcal{A}_\infty(\eta)$ and β a C^∞-path of C^∞-forms in $\Omega^0(N)$. By hypothesis the path $B = B(t)$ lies in

$$\mathcal{G}_{\frac{5}{2}}(\eta) \times_{Stab(\Gamma)} U_\Gamma.$$

Consider the subset Σ of $\mathcal{G}_{\frac{5}{2}}(\eta) \times I$ consisting of all pairs $\{(\sigma,t)|\sigma^*B(t) \in U_\Gamma\}$. If $(\sigma,t) \in \Sigma$, then $\sigma^*B(t) \in U_\Gamma$ and hence by Remark 2.3.3 $\sigma^*B(t)$ is C^∞ and hence so is σ. We claim that Σ is a smooth $Stab(\Gamma)$-subbundle of the product. Given this, there will be a smooth section $t \mapsto \sigma(t)$ which forms a C^∞-gauge transformation σ on $I \times N$. Clearly, $\sigma^*\omega = A(t) + \alpha(t)dt$ as required.

It remains to prove that Σ is a smooth $Stab(\Gamma)$ subbundle of $\mathcal{G}_{\frac{5}{2}} \times I$. For this it suffices to find local C^∞-sections $t \mapsto \sigma(t)$. Once we are working locally in t, it suffices to assume that $B(t)$ is close to U_Γ. With this assumption, the implicit function theorem tells us that there is a unique gauge transformation σ_0 carrying $B(t)$ into U_Γ which in addition is the image under the exponential map of a small element in $(Ker\Delta_\Gamma)^\perp \subset \Omega^0(N)$ and that σ_0 varies smoothly with $B(t)$. □

We now prove Lemma 2.4.3.

Proof. By Lemma 2.4.5 we can assume that

$$\omega = B(t) + \beta(t)dt$$

where $B(t)$ is a C^∞-path of C^∞-connections in U_Γ and $\beta(t)$ is a C^∞-path of C^∞ forms in $\Omega^0(N)$. Let $\beta^0(t)$ denote the projection of $\beta(t)$ onto $Ker(\Delta_\Gamma)$, and define the gauge transformation σ to be the unique C^∞ solution to the initial value problem

$$\frac{\partial \sigma}{\partial t} = -\beta^0(t)\sigma \text{ and } \sigma(0) = Id.$$

Clearly such a σ has $\sigma(t) \in Stab(\Gamma)$. From the equation

$$\sigma^*(\omega) = B + \beta dt + \sigma^{-1}\left(d_B\sigma + \left(\frac{\partial \sigma}{\partial t} + [\beta, \sigma]\right) dt\right)$$

we have $\sigma^*\omega = A(t) + \alpha(t)dt$ where $\alpha = \sigma^{-1}(\beta - \beta^0)\sigma \in Ker(\Delta_\Gamma)^\perp$.

Now let us establish that σ is unique up to composition with a constant gauge transformation in $Stab(\Gamma)$. Suppose that $\sigma_0^*\omega = A(t) + \alpha(t)dt$ and $\sigma_1^*\omega = B(t) + \beta(t)dt$ are both in standard form with respect to U_Γ. Consider $\tau(t) = \sigma_0(t)\sigma_1^{-1}(t)$. This means that

$$\tau(t)^* B(t) = A(t) \text{ and} \tag{7}$$

$$\tau(t)^{-1}\beta(t)\tau(t) + \tau(t)^{-1}(\partial\tau)/(\partial t) = \alpha(t). \tag{8}$$

Equation 7 implies that $\tau(t) \in Stab(\Gamma)$ for all t. Thus, conjugation by $\tau(t)$ preserves the subspace $(Ker\Delta_\Gamma)^\perp$. Hence from Equation 8 we see that

$$\tau(t)^{-1}\frac{\partial \tau}{\partial t} \in (Ker\Delta_\Gamma)^\perp.$$

Since $\tau(t) \in Stab(\Gamma)$ for all t, this implies that $\tau(t)$ is a constant function of t. □

In the appendix to this chapter we give a version of Lemma 2.4.3 for Sobolev connections.

2.5. The restriction of the Chern-Simons function to a slice through a flat connection

For all of this section we fix a flat C^∞-connection Γ on η. Let us consider $f_\Gamma = CS|S_\Gamma$, the Chern-Simons function restricted to the slice S_Γ. The purpose of this section is to compute the gradient $\nabla_A f_\Gamma$ of f_Γ at A **with respect to the inner product** $\langle\cdot,\cdot\rangle_{\{\Gamma,A\}}$. In general if $A \in U_\Gamma$, then $\nabla_A CS$ will not be tangent to U_Γ, and hence, $\nabla_A f_\Gamma$ will not be equal to $\nabla_A CS$, even ignoring the change in the inner product. As the next lemma shows, the difference of $\nabla_A f_\Gamma$ and $\nabla_A CS$ is measured by $d_A\Theta(A)$ for an appropriate function $\Theta\colon U_\Gamma \to \Omega^0_2(N)$. For our later applications it will be important to have several estimates concerning Θ. The following lemma gives some of these. Others are given in the next section.

LEMMA 2.5.1. *There is a coordinate patch U_Γ centered at Γ and a unique smooth $Stab(\Gamma)$-equivariant map $\Theta\colon U_\Gamma \to \Omega^0_2(N)$ such that for all $A \in U_\Gamma$ we have*

$$\begin{aligned} d^*_\Gamma(*F(A) - d_A(\Theta(A))) &= 0 \quad \textit{and} \\ \Theta(A) &\in (Ker\Delta_\Gamma)^\perp. \end{aligned}$$

Furthermore, the following hold:

(1) *For $A \in U_\Gamma$, say $A = \Gamma + a$, we have the following equation in $\Omega^1_{\frac{1}{2}}(N)$*

$$\nabla_A f_\Gamma = -*F_A + d_A(\Theta(A)) \tag{9}$$

in the sense that $*F_A - d_A(\Theta(A)) \in Ker(d^*_\Gamma)$ *and for all* $v \in (TU_\Gamma)_A$ *we have*

$$D_A f_\Gamma(v) = \langle - * F_A + d_A(\Theta(A)), v\rangle_{\{\Gamma,A\}}. \tag{10}$$

(2) *There is a constant* $K_0 > 0$ *depending on* U_Γ *such that for all* $A \in U_\Gamma$ *we have:*

$$||\Theta(A)||_{L^2_{\frac{7}{4}}} \leq K_0||a||_{L^2_{\frac{3}{2}}}||\nabla_A f_\Gamma||_{L^2} \tag{11}$$

$$||\Theta(A)||_{C^0} \leq K_0||a||_{L^2_{\frac{3}{2}}}||\nabla_A f_\Gamma||_{L^2} \tag{12}$$

Proof. Recall that we denote $A-\Gamma$ by a. Then $d^*_\Gamma(*F_A) = - * [d_\Gamma a, a]$. By the continuous multiplication $L^2_{\frac{1}{2}} \otimes L^2_{\frac{3}{2}} \to L^2$ we see that $-d^*_\Gamma(*F_A)$ is L^2. For A in $\mathcal{A}_{\frac{3}{2}}(\eta)$ and any $r \leq 2$, the following is a bounded operator depending analytically on A:

$$d^*_\Gamma \circ d_A : \Omega^0_r(N) \cap Ker(\Delta_\Gamma)^\perp \to \Omega^0_{r-2}(N) \cap Imd^*_\Gamma.$$

Since it is an isomorphism for $A = \Gamma$, for all A in a sufficiently small U_Γ this operator is an isomorphism whose inverse is uniformly bounded. Since the map $\Omega^1_{\frac{3}{2}}(N) \to \Omega^0_0(N)$ given by $A \mapsto d^*_\Gamma(*F_A)$ is smooth, the existence, uniqueness and smoothness of $\Theta(A)$ satisfying the first two equations is now immediate.

Let us prove Part 1. First notice that $- * F_A + d_A(\Theta(A))$ lies in $\mathcal{K}_\Gamma$ and its image modulo $d_A(Ker(\Delta_\Gamma)^\perp)$ agrees with $\nabla_A CS = - * F_A$. We claim that Equation 10 follows from these two properties. To see this let v be an element in $(TU_\Gamma)_A = \mathcal{K}_\Gamma$. Then

$$D_A f_\Gamma(v) = \langle \nabla_A CS, v\rangle_{L^2} = \langle - * F_A, v\rangle_{L^2}.$$

Since $*F_A \in Kerd^*_A$ and $v - p_A(v) \in Imd_A$ we have

$$\langle - * F_A, v\rangle_{L^2} = \langle - * F_A, p_A(v)\rangle_{L^2}.$$

Since $p_A(v) \in \mathcal{K}'_A$ and $\Theta \in (Ker(\Delta_\Gamma)^\perp)$ we also have

$$\langle - * F_A, p_A(v)\rangle_{L^2} = \langle - * F_A + d_A(\Theta(A)), p_A(v)\rangle_{L^2}.$$

Clearly,

$$\begin{aligned}\langle - * F_A + d_A(\Theta(A)), p_A(v)\rangle_{L^2} &= \langle p_A(- * F_A + d_A(\Theta(A))), p_A(v)\rangle_{L^2} \\ &= \langle - * F_A + d_A(\Theta(A)), v\rangle_{\{\Gamma,A\}}.\end{aligned}$$

This proves the formula in Part 1 for $\nabla_A f_\Gamma$.

As to Part 2 we have

$$d_\Gamma^* d_A\,(\Theta(A)) = d_\Gamma^*(*F_A) = - * d_\Gamma F_A = *[a, F_A] \tag{13}$$

Using the continuous multiplication $L^2_{\frac{3}{2}} \otimes L^2 \to L^{\frac{7}{4}}$ we see that for some constant C_0 we have

$$||d_\Gamma^* d_A(\Theta(A))||_{L^{\frac{7}{4}}} \leq C_0 ||a||_{L^2_{\frac{3}{2}}} ||F_A||_{L^2}.$$

By the definition of $\Theta(A)$ this implies that

$$||d_\Gamma^* d_A(\Theta(A))||_{L^{\frac{7}{4}}} \leq C_0 ||a||_{L^2_{\frac{3}{2}}} \left(||\nabla_A(f_\Gamma)||_{L^2} + ||d_A\Theta(A)||_{L^2}\right). \tag{14}$$

From the definition of d_A we have

$$||d_A(\Theta(A))||_{L^2} \leq ||d_\Gamma(\Theta(A))||_{L^2} + ||[a, \Theta(A)]||_{L^2}.$$

Using the continuous embedding $L^2_{\frac{7}{4}} \subset L^2_1$ and the continuous multiplication $L^2_{\frac{3}{2}} \otimes L^2_{\frac{7}{4}} \to L^2$, this yields

$$||d_A(\Theta(A))||_{L^2} \leq C_0' ||\Theta(A)||_{L^2_{\frac{7}{4}}} + C_0'' ||a||_{L^2_{\frac{3}{2}}} ||\Theta(A)||_{L^2_{\frac{7}{4}}} \tag{15}$$

for some constants C_0' and C_0''. Inequalities 14 and 15 give:

(16)

$$||d_\Gamma^* d_A(\Theta(A))||_{L^{\frac{7}{4}}} \leq C_0 ||a||_{L^2_{\frac{3}{2}}} \left(||\nabla_A(f_\Gamma)||_{L^2} + (C_0' + C_0'' ||a||_{L^2_{\frac{3}{2}}}) ||\Theta(A)||_{L^2_{\frac{7}{4}}}\right).$$

For A sufficiently close in $L^2_{\frac{3}{2}}$ to Γ the operators $d_\Gamma^* d_A \colon (Ker\Delta_\Gamma)^\perp \cap \Omega^0_{L^{\frac{7}{4}}_2} \to Im d_\Gamma^* \cap \Omega^0_{L^{\frac{7}{4}}}$ are uniformly invertible. Thus, there is a constant C_1 such that

$$||\Theta(A)||_{L^{\frac{7}{4}}_2} \leq C_1 ||d_\Gamma^* d_A(\Theta(A))||_{L^{\frac{7}{4}}}.$$

Together with Inequality 16 this gives

$$||\Theta(A)||_{L^{\frac{7}{4}}_2} \leq C_2||a||_{L^2_{\frac{3}{2}}} \left(||\nabla_A(f_\Gamma)||_{L^2} + (1 + C_0'||a||_{L^2_{\frac{3}{2}}})||\Theta(A)||_{L^2_{\frac{7}{4}}}\right)$$

for some constant C_2. Since $L^{\frac{7}{4}}_2$ embeds continuously in $L^2_{\frac{7}{4}}$ there is a constant C_3 such that

$$||\Theta(A)||_{L^2_{\frac{7}{4}}} \leq C_3||a||_{L^2_{\frac{3}{2}}} \left(||\nabla_A(f_\Gamma)||_{L^2} + (1 + C_0'||a||_{L^2_{\frac{3}{2}}})||\Theta(A)||_{L^2_{\frac{7}{4}}}\right).$$

For U_Γ sufficiently small, Inequality 11 follows.

Inequality 12 follows immediately from Inequality 11 and the continuous embedding $L^2_{\frac{7}{4}}(N) \subset C^0(N)$. □

Notice that Inequality 11 says that for A sufficiently close to Γ the projection of $\nabla_A(CS)$ onto either $Im\, d_\Gamma$ or $(\mathcal{K}'_A)^\perp$ has small norm compared with the norm of $\nabla_A f_\Gamma$.

COROLLARY 2.5.2. *There is a coordinate patch U_Γ such that for all $A, A' \in U_\Gamma$ we have:*

$$\frac{3}{4}||\nabla_A f_\Gamma||_{L^2} \leq ||F_A||_{L^2} \leq \frac{5}{4}||\nabla_A(f_\Gamma)||_{L^2} \tag{17}$$

Proof. This is immediate from Inequalities 11 and 9. □

2.5.1. Estimates on the nonlinearity of ∇f_Γ. We shall need estimates similar to the ones we established in Lemma 2.5.1 for the function

$$n(a) = \nabla_A f_\Gamma + *d_\Gamma(a) = - * [a,a] + d_{\Gamma+a}(\Theta(\Gamma + a)).$$

Since $-*d_\Gamma$ is the linearization of $\nabla_A f_\Gamma$ at A, this function measures the nonlinearity of $\nabla_A f_\Gamma$ at A.

LEMMA 2.5.3. (1) *Let $U \subset \mathcal{K}_\Gamma$ be such that $\{\Gamma\} + U = U_\Gamma$ satisfies the conclusions of Lemma 2.5.1. For any $k > \frac{3}{2}$ the map n is a smooth map from $U \cap L^2_k$ to $\Omega^1_k(N)$.*

(2) *There is a constant C_1 (depending on U_Γ) such that for any $a, a' \in U \cap L^2_2$ and any $b \in \mathcal{K}_\Gamma \cap L^2_2$ we have:*

$$\|D_a n(b)\|_{L^2} \leq C_1\|a\|_{L^2_2}\|b\|_{L^2} \tag{18}$$

$$\|n(a) - n(a')\|_{L^2} \leq C_1\left(\|a\|_{L^2_2} + \|a'\|_{L^2_2}\right)\|a - a'\|_{L^2} \tag{19}$$

In particular, the map $n\colon U \cap L^2_2 \to \Omega^1_2(N)$ vanishes to second order at $\{0\}$.

Proof. To prove Part 1 fix $k > (3/2)$. For $A \in U_\Gamma \cap L^2_k$ the operator

$$d^*_\Gamma d_A\colon \Omega^0_{k+1}(N) \to \Omega^0_{k-1}(N)$$

is a smooth isomorphism which depends smoothly on A. Hence so does its inverse. This, together with Equation 13 and the usual Sobolev multiplication theorem, implies that $\Theta\colon U_\Gamma \cap L^2_k \to \Omega^0_{k+1}(N)$ is a smooth map. It follows immediately from this and the Sobolev multiplication theorem that $n\colon U \cap L^2_k \to \Omega^1_k(N)$ is a smooth map.

We consider the first inequality in Part 2. The differential of n is given by

$$D_a n(b) = -2 * a \wedge b + D_a\left\{d_{\Gamma+a}\Theta(\Gamma + a)\right\}(b).$$

By the Sobolev multiplication theorem the first term clearly satisfies the desired inequality. We consider $D_a\left\{d_{\Gamma+a}\Theta(\Gamma + a)\right\}(b)$. We have

$$D_a\left\{d_{\Gamma+a}\Theta(\Gamma + a)\right\}(b) = d_{\Gamma+a}(D_A\Theta)(b) + [b, \Theta(A)].$$

Since Θ is a smooth function from U_Γ to $\Omega^0_2(N)$ with $\Theta(\Gamma) = 0$, if we choose U_Γ sufficiently small, then there is a constant K such that $\|\Theta(A)\|_{L^2_2} \leq K\|a\|_{L^2_{\frac{3}{2}}}$. Thus, with this choice of U_Γ there is a constant C_0 such that

$$\|[b, \Theta(A)]\|_{L^2} \leq C_0\|b\|_{L^2} \cdot \|a\|_{L^2_{\frac{3}{2}}}. \tag{20}$$

Thus, we need only see that there is a constant C_1' such that

$$\|d_{\Gamma+a}(D_A\Theta)(b)\|_{L^2} \leq C_1'\|a\|_{L^2_2} \cdot \|b\|_{L^2}.$$

Differentiating Equation 13, we see

$$\begin{aligned} d_\Gamma^* D_a \{d_{\Gamma+a}\Theta\}(b) &= D_a \{d_\Gamma^* d_{\Gamma+a}\Theta\}(b) \\ &= D_a \{- * d_\Gamma F_A\}(b) = - * d_\Gamma[a,b]. \end{aligned}$$

Hence, by the Sobolev multiplication theorem there is a constant C' such that

$$\|d_\Gamma^* D_a \{d_{\Gamma+a}\Theta\}(b)\|_{L^2_{-1}} \leq C'\|a\|_{L^2_2}\|b\|_{L^2}. \tag{21}$$

But

$$d_\Gamma^* D_a \{d_{\Gamma+a} D_{\Gamma+a}\Theta\}(b) = d_\Gamma^* d_{\Gamma+a}\left((D_A\Theta)(b)\right) + d_\Gamma^*[b, \Theta(A)].$$

By Inequalities 20 and 21 this implies that there is a constant C_1'' with

$$\|d_\Gamma^* d_{\Gamma+a}\left(D_A(\Theta)(b)\right)\|_{L^2_{-1}} \leq C_1''\|a\|_{L^2_2} \cdot \|b\|_{L^2}.$$

Since $d_\Gamma^* d_{\Gamma+a}$ is uniformly invertible as an operator from L^2_1 to L^2_{-1} it follows that there is a constant C_2' with

$$\|(D_A\Theta)(b)\|_{L^2_1} \leq C_2'\|a\|_{L^2_2} \cdot \|b\|_{L^2}$$

and hence there is a constant C_1'

$$\|d_{\Gamma+a}(D_A\Theta)(b)\|_{L^2} \leq C_1'\|a\|_{L^2_2} \cdot \|b\|_{L^2}.$$

This completes the proof of the first inequality in Part 2. The second inequality in Part 2 is immediate from the first and the mean value theorem. □

2.6. The flow equation on a slice determined by the ASD Equation

We wish to study the nature of an ASD connection ω on $[0,\infty) \times \eta$ whose path $A(t) = \omega|\{t\} \times \eta$ up to gauge transformation lies in a coordinate patch U_Γ centered at Γ. We shall be interested in perturbations of the product metric, but let us start with the case of the product metric. This case is clear from the work of the Section 2.5.

LEMMA 2.6.1. *Let $I \subset [0,\infty)$ be an interval. If $\omega = A(t) + \alpha(t)dt$ is an g_0-ASD connection on $I \times \eta$ in standard form with respect to Γ, then the path $A(t)$ satisfies the gradient flow equation for f_Γ, i.e.,*

$$\frac{\partial A(t)}{\partial t} = \nabla_{A(t)} f_\Gamma$$

and $\alpha(t) = \Theta(A(t))$.

Proof. Clearly, $d_\Gamma^*(\partial A/\partial t) = 0$. Applying d_Γ^* to Equation 3 it follows that $d_\Gamma^*(*F_A) = d_\Gamma^* d_A(\alpha)$. Thus, by the uniqueness statement in Lemma 2.5.1, since $\alpha \in Ker(\Delta_\Gamma)^\perp$ we have $\alpha = \Theta(A)$. The first part of the result now follows from this and Equations 3 and 9. □

Next we show that if $\omega|\{t\} \times \eta$ is conjugate into U_Γ for all $t \in I$ then the connection ω can be conjugated by a gauge transformation so that it is in standard from.

LEMMA 2.6.2. *Let U_Γ be as in Lemma 2.5.1 and let $I \subset [0,\infty)$ be an interval. Let ω be an g_0-ASD connection on*

$$I \times \eta$$

which is L_2^2. Suppose for each t that $\omega|(\{t\} \times \eta)$ is $L^2_{\frac{5}{2}}$-gauge equivalent to a connection in U_Γ. Then there is a L_3^2-gauge transformation $\sigma = \sigma(t)$ and a C^∞-path of C^∞-connections $A = A(t)$, $t \in I$, in U_Γ satisfying the gradient flow equation

$$\frac{\partial A(t)}{\partial t} = \nabla_{A(t)} f_\Gamma \tag{22}$$

such that

$$\sigma^*\omega(t) = A(t) + \Theta(A(t))dt.$$

The gauge transformation σ is unique up to composition with with the pullback of an automorphism $\tau_0 \in Stab(\Gamma)$.

Proof. Since ω is ASD with respect to a C^∞- metric, there is a L^2_3-gauge transformation σ_0 with $\sigma_0^*\omega$ a C^∞-connection. By Lemma 2.4.3 there is a further C^∞-gauge transformation σ_1 such that $\sigma_1^*\sigma_0^*\omega = A(t) + \alpha(t)dt$ is in standard form. The result is now immediate by Lemma 2.6.1. □

So far we have been dealing with the product metric $g_0 = g_N + dt^2$. Now we are ready to study the flow equation in a slice for a metric g such that $g - g_0$ bounded in the C^3 topology. The result for these metrics is similar to Lemma 2.6.2 except that the flow equation is the gradient of f_Γ plus an error term with explicit bounds. This is the main technical result that allows us to apply Simon's theorems to our situation for non-product metrics.

THEOREM 2.6.3. *There is a neighborhood B of the product metric g_0 in the space of C^3-metrics with the uniform C^3-topology such that for any C^∞ flat connection Γ there is a coordinate patch U_Γ centered at Γ such that the following holds. Let $g \in B$. Suppose that $1 \le T_0 < T_1 \le \infty$. Suppose that ω is a g-ASD connection on $(T_0 - 1, T_1 + 1) \times \eta$ which is $L^2_{2,loc}$ such that for all t, $T_0 \le t < T_1$, the connection $\omega|(\{t\} \times \eta)$ is $L^2_{\frac{5}{2}}$-gauge equivalent to a connection in U_Γ. Then there is an $L^2_{3,loc}$-gauge transformation $\sigma = \sigma(t)$ on $[T_0, T_1) \times \eta$, unique up to composition with the pullback of an automorphism $\tau_0 \in Stab(\Gamma)$ such that $\sigma^*\omega = A + \alpha dt$ is a C^3-connection in standard form with respect to U_Γ. In addition, defining the error term $\wp = \wp(t)$ for A and α by the equation*

$$\frac{\partial A(t)}{\partial t} = \nabla_{A(t)} f_\Gamma + \wp(t) \tag{23}$$

we have that $\wp$ is a C^2 one-form on $[T_0, T_1) \times N$. Furthermore, the following estimates hold for $t \in [T_0, T_1)$:

$$\|\alpha(t)\|_{L^2_1} \le \frac{1}{8}\|\nabla_{A(t)} f_\Gamma\|_{L^2} \tag{24}$$

$$\|\wp(t)\|_{L^2} \le \frac{1}{2}\|\nabla_{A(t)} f_\Gamma\|_{L^2} \tag{25}$$

Also there are constants K_0 *and* K_1 *depending only on* U_Γ *so that*

$$\|\wp(t)\|_{L^2} < K_0\|g(t) - g_0(t)\|_{C^0}\|\nabla_{A(t)}f_\Gamma\|_{L^2} \tag{26}$$

and

$$\|\alpha(t) - \Theta(A(t))\|_{C^0} \leq K_1\|g(t) - g_0(t)\|_{C^1}\|F_A\|_{L^2_1}. \tag{27}$$

Note: For $0 \leq i \leq 3$, the paths $A - \Gamma$ and α are C^i-paths into the space of C^{3-i} forms on N.

The proof of this theorem appears in the next chapter.

COROLLARY 2.6.4. *If* ω_1 *and* ω_2 *are* C^3 *g-ASD connections in standard form on* $(T_0 - 1, T_1 + 1) \times \eta$ *and if* σ *is a gauge transformation carrying* ω_1 *to* ω_2*, then* $\sigma|_{[T_0,T_1)\times\eta}$ *is the pullback of a connection* $\tau_0 \in Stab(\Gamma)$.

DEFINITION 2.6.5. *A coordinate patch* U_Γ *centered at* Γ *for which Theorem* 2.6.3 *holds is said to be a* gradient flow patch.

CHAPTER 3

Regularity and Estimates for the Flow Equation: The Proof of Theorem 2.6.3

Theorem 2.6.3 is proved as follows. First we show that certain kinds of weak Sobolev connections can be put in standard form. Then we show that for g-ASD connections in standard form the desired estimates of Theorem 2.6.3 hold. Finally we show that g-ASD connections have the regularity claimed in the theorem. The chapter ends with estimates which will be used later.

3.1. Putting Sobolev connections in standard form

Our first discussion generalizes Lemma 2.4.3 and shows how to put a certain type of Sobolev connection in standard form.

As in the C^∞ situation we will use the parameterized version of the implicit function theorem to put our connection into standard form. The notion of standard form involves only a condition along slices $\{t\} \times N$ and hence the gauge transformation so obtained has better Sobolev regularity (by one derivative) than the connection along the slice $\{t\} \times N$ while in the t direction it has the same regularity as the connection. This forces us to use Sobolev spaces with different regularity in the t direction than along $\{t\} \times N$. In the next section we establish a regularity theorem for ASD connections in standard form. Using this we can recover the lost regularity and return to working with more standard Sobolev spaces.

We begin by recalling the basic definitions and properties of anisotropic Sobolev spaces. More details on this material can be found in Hörmander [16, Vol. III appendix B]. Fix for the moment nonnegative integers m, s.

DEFINITION 3.1.1. *Let $L^2_{(m,s)}([T_0, T_1] \times N)$ denote the set of functions $f \in L^2([T_0, T_1] \times N)$ so that for all i, j with $i + j \leq m + s$ and $i \leq m$ we have*

$$\frac{\partial^i}{\partial t^i} \nabla^{(j)} f \in L^2([T_0, T_1] \times N).$$

For $f \in L^2_{(m,s)}([T_0, T_1] \times N)$ define the (m, s) norm of f to be

$$||f||^2_{L^2_{(m,s)}} = \sum_{i+j \leq m+s, i \leq m} ||\frac{\partial^i}{\partial t^i} \nabla^{(j)} f||_{L^2}.$$

Thus $L^2_{(m,s)}$ is an analogue of the Sobolev space L^2_{m+s} except that one does not allow t derivatives of order higher than m.

As usual there is a similar definition for sections of a vector bundle. If P denotes the product bundle $[T_0, T_1] \times \eta$ the keeping with our previous notation $\Omega^k_{(m,s)}([T_0, T_1] \times N)$ will denote the space of $L^2_{(m,s)}$ k-forms with values in $Ad\ P$. These spaces enjoy various embedding and multiplication properties, analogous to those for the usual Sobolev spaces. These are summarized below.

THEOREM 3.1.2. *(1) $L^2_{(m,s)} \hookrightarrow L^2_{(m',s')}$ whenever $m' \leq m$ and $m' + s' \leq m + s$. This inclusion is compact whenever both inequalities are strict.*

(2) $L^2_{(m,s)} \hookrightarrow C^k$ whenever $k + n/2 < m + s$ and $k + 1/2 < m$.

(3) The multiplication $L^2_{(m',s')} \otimes L^2_{(m,s)} \to L^2_{(m,s)}$ is continuous if $n/2 < m' + s'$, $1/2 < m'$, $m \leq m'$, and $m + s \leq m' + s'$.

(4) The restriction $L^2_{(m,s)} \to L^2_{m+s-1/2}(\{t\} \times N)$ is continuous if $m > 1/2$

Next we introduce the space of connections and the group of gauge transformations of interest to us. Let $\mathcal{G}_{(1,2)}$ denote the group of $L^2_{(1,2)}$ gauge transformations of P. Also let

$$\mathcal{A}_{(1,1)} = \{\Gamma + a + \alpha dt | a \in L^2_{(1,1)}([T_0, T_1] \times N; \pi^* \Lambda^1 N \otimes AdP),\ \alpha \in \Omega^0_{(0,2)}([T_0, T_1] \times N)\}.$$

Note that for the space of connections dt and the T^*N components of the connection have different Sobolev regularity.

Let us make a few remarks about these spaces. Theorem 3.1.2 implies $L^2_{(1,2)} \hookrightarrow C^0$ and that the multiplications $L^2_{(1,2)} \otimes L^2_{(1,1)} \to L^2_{(1,1)}$ and $L^2_{(1,2)} \otimes L^2_{(0,2)} \to L^2_{(0,2)}$ are continuous. Also note that the exterior derivative along N defines a continuous map $d\colon L^2_{(1,2)} \to L^2_{(1,1)}$ while the time derivative defines a continuous map $\frac{\partial}{\partial t}\colon L^2_{(1,2)} \to L^2_{(0,2)}$. Given the above remarks the proof in [12] appendix B adapts word for word to show that $\mathcal{G}_{(1,2)}$ is a Hilbert Lie group which acts smoothly on $\mathcal{A}_{(1,1)}$ by gauge transformation.

The main result of this section is the following slice lemma.

LEMMA 3.1.3. *Let Γ be a C^∞ flat connection. Let U_Γ be a coordinate patch centered at Γ and let $I \subset [0,\infty)$ be an interval. Suppose that ω is an $L^2_{(1,1)}$ connection on $I \times \eta$ such that $\omega|_{\{t\}\times n}$ is $L^2_{\frac{5}{2}}$ gauge equivalent to a connection in U_Γ for all $t \in I$. Then there is a $\sigma \in \mathcal{G}_{(1,2)}(I \times \eta)$ so that $\sigma^*(\omega) = A + \alpha dt$ is in standard form with respect to U_Γ*

The gauge transformation σ is unique up to composition with the pullback of an element of $Stab(\Gamma)$.

The uniqueness statement has the following useful corollary.

COROLLARY 3.1.4. *Let ω_1, ω_2 be $L^2_{(1,1)}$ connections on $I \times \eta$ which are in standard form with respect to U_Γ. If there is a gauge transformation σ so that $\sigma^*(\omega_1) = \omega_2$ then σ is the pullback of an element of $Stab(\Gamma)$.*

Before proving this it is convenient, just as in the C^∞ case, to first deal with the T^*N component of the connection.

LEMMA 3.1.5. *For Γ, U_Γ, and ω as in the hypothesis of Lemma 3.1.3 there is an $L^2_{(1,2)}$ gauge transformation σ so that $\sigma^*\omega = A(t) + \alpha(t)dt$ where $A(t) \in U_\Gamma$.*

Proof. The crucial claim is the following.

CLAIM 3.1.6. *Fix $\omega = B + \beta dt \in \mathcal{A}_{(1,1)}$ as in Lemma 3.1.3, and $t_0 \in I$. There is a neighborhood I' containing t and gauge transformation $\sigma \in \mathcal{G}_{(1,2)}(I' \times N)$ so that $\sigma^*\omega|_{\{t\}\times N} \in U_\Gamma$ for all $t \in I'$.*

Proof. For the proof we need the L_1^2 version of the coordinate patch U_Γ. Namely, set U_1 equal to the interior of the L_1^2 closure of U_Γ. By assumption there is a gauge transformation $\tau \in \mathcal{G}_{\frac{5}{2}}$ so that $\tau^* B(t_0) \in U_\Gamma$. The extension theorem (cf [16]) guarantees that on a neighborhood $I'' \times \eta$ of $\{t\} \times \eta$, there is an L_3^2 gauge transformation τ_1 so that $\tau_1|_{\{t\}\times\eta} = \tau$. Hence we may assume that $B(t_0) \in U_\Gamma$. Since $\omega \in \mathcal{A}_{(1,1)}$, it can be viewed both as an L^2 path of connections in $\mathcal{A}_2$ and an L_1^2 path of connections in $\mathcal{A}_1$.

For any $k \geq 1$, the map $\Phi_k : U_1 \cap L_k^2 \times Ker(\Delta_\Gamma)^\perp \cap L_{k+1}^2 \to \mathcal{A}_k(\eta)$ given by the local action of the gauge algebra is a local diffeomorphism. The fact that Φ_1 and Φ_2 are local diffeomorphisms implies the following. There are unique paths τ_2 and A defined on a subinterval $I' \subset I''$ containing t, such that τ_2 is simultaneously an L^2 path of L_3^2 gauge transformations and an L_1^2 path of L_2^2 gauge transformations, and $\tau_2^*\omega|_{\{t\}\times\eta} = A(t)$ is simultaneously an L^2 path of connections in $U_\Gamma \cap L_2^2$ and an L_1^2 path of L_1^2-connections. It follows that $\tau_2 \in \mathcal{G}_{(1,2)}(I' \times \eta)$. □

Given this claim the proof of Lemma 3.1.5 is as follows. Let $I' \subset I$ denote the maximal subinterval containing the initial point of I so that there is a gauge transformation σ of $I' \times \eta$ so that Lemma 3.1.5 holds. Clearly I' is not empty, as $T_0 \in I$. For ω satisfying the hypothesis of Lemma 3.1.3 the conclusion of Lemma 3.1.5 is a closed condition in t hence I' is closed in I. Finally the previous claim guarantees that I' is open in I. Hence, $I' = I$.

The proof that a gauge transformation exists so that part (b) of the definition of standard form holds, and the proof of the uniqueness assertion are the same as for the C^∞ case. □

The proof of Lemma 3.1.3 has the following useful corollary.

COROLLARY 3.1.7. *There are constants $C, \epsilon > 0$ so that the following holds. Let $\omega \in \mathcal{A}_{(1,1)}(I \times \eta)$ be a connection so that $\iota = \{\inf ||\omega - \omega'||_{\mathcal{A}_{(1,1)}} | \omega' \text{ in standard form}\} < \epsilon$. Then there is a $\sigma \in \mathcal{G}_{(1,2)}(I \times \eta)$ so that $\sigma^*(\omega)$ is in standard form with respect to Γ and*

$$||\sigma - Id||_{L^2_{(1,2)}} \leq C\iota. \tag{28}$$

3.2. Estimates for g-ASD connections in standard form

As a first step toward the proof of Theorem 2.6.3 let us show that the desired estimates follow for g-ASD Sobolev connections which are in standard form.

LEMMA 3.2.1. *Let Γ be a C^∞ flat connection Γ. There is a coordinate patch U_Γ centered at Γ and constants $b, C_0, C_1 > 0$ so that the following holds. Let g be a C^3 metric of the form $g = g_0 + h$ with $||h||_{C^3} < b$. Suppose that $1 \leq T_0 < T_1 \leq \infty$. Suppose that $\omega = A + \alpha dt$ is a g-ASD connection on $[T_0 - 1, T_1 + 1) \times \eta$ which is in the anisotropic Sobolev space of connections $\mathcal{A}_{(1,1)}([T_0 - 1, T_1 + 1) \times \eta)$, and that ω is in standard form with respect to U_Γ.*

Then defining the error term $\wp = \wp(t)$ for A and α by the equation

$$\frac{\partial A(t)}{\partial t} = \nabla_{A(t)} f_\Gamma + \wp(t) \tag{29}$$

we have that $\wp$ is in $\Omega^1_{(1,0)}([T_0 - 1, T_1 + 1) \times \eta)$ and for all t in $[T_0 - 1, T_1 + 1]$ we have that $\wp(t) \in Ker d^*_\Gamma \subset \Omega^1_1(N)$

Furthermore, the following estimates hold for $t \in [T_0, T_1]$,

$$||\alpha(t)||_{L^2_1} \leq \frac{1}{8}||\nabla_A f_\Gamma||_{L^2} \tag{30}$$

$$||\wp(t)||_{L^2} < C_0 ||h(t)||_{C^0} ||\nabla_{A(t)} f_\Gamma||_{L^2}. \tag{31}$$

Lastly, for all $t \in [T_0 - 1, T_1 + 1)$ so that $F_A \in L^2_1$ we have

$$||\alpha(t) - \Theta(A(t))||_{C^0} \leq C_1 ||h(t)||_{C^1} ||F_{A(t)}||_{L^2_1} \tag{32}$$

Proof. We fix U_Γ such that Lemma 2.5.1 and Corollary 2.5.2 hold. Fix $b > 0$ such that Lemma 2.2.1 holds. Let g be a C^3-metric with $||g - g_0||_{C^3} < b$, and let ω be a g-ASD connection satisfying the hypothesis of the theorem with respect to U_Γ.

It follows from Equation 5 and the fact that $A \in S_\Gamma$ and hence that $\frac{\partial A}{\partial t} \in \mathcal{K}_\Gamma$ that we have

$$d_\Gamma^* d_A \alpha = d_\Gamma^* \phi_h(*F_A). \tag{33}$$

Thus, we have

$$\begin{aligned} d_\Gamma^* d_\Gamma \alpha &= d_\Gamma^*([a,\alpha]) - *d_\Gamma * (\phi_h(*F_A)) \\ &= d_\Gamma^*([a,\alpha]) - * \bigwedge((Id \otimes (*\phi_h *))(\nabla_\Gamma F_A)) - *[\nabla\phi_h, (F_A)] \end{aligned}$$

This last equation requires some interpretation. $\bigwedge\colon \Lambda^1(Ad\,\eta) \otimes \Lambda^2(Ad\,\eta) \to \Lambda^3(Ad\,\eta)$ is the antisymmetrization. $\nabla_\Gamma F_A$ denotes the full covariant derivative of the two form F_A and hence is a section of $\Omega^1(Ad\,\eta) \otimes \Omega^1(Ad\,\eta)$. Finally $\nabla(\phi_h)$ is a one form with values in the endomorphisms of the one forms obtained by taking the covariant derivative of ϕ_h with respect to the Levi-Civita connection. With this notation the Bianchi identity takes the form $\bigwedge(\nabla_A F_A) = \bigwedge(\nabla_\Gamma F_A + a \otimes F_A) = 0$. Combining this with the last equation gives

(34)

$$d_\Gamma^* d_\Gamma \alpha = d_\Gamma^*([a,\alpha]) - * \bigwedge((Id \otimes (*\phi_h * - Id))(\nabla_\Gamma F_A) + (a \otimes F_A)) - *[\nabla\phi_h, (F_A)]$$

Using the continuity of the multiplication $L^2_{\frac{3}{2}} \otimes L^2_k \to L^2_{k-1}$ for $k < \frac{3}{2}$ and the fact that $\Delta_\Gamma\colon Ker(\Delta_\Gamma)^\perp \cap L^2_1 \to Ker(\Delta_\Gamma)^\perp \cap L^2_{-1}$ is an isomorphism, it follows from Equation 34 that there is a constant $C > 0$ such that

$$\begin{aligned} ||\alpha||_{L^2_1} \;\le\; C \Big(& ||a||_{L^2_{\frac{3}{2}}} ||\alpha||_{L^2_1} + ||\phi_h - Id||_{C^0} ||F_A||_{L^2} \\ & + ||a||_{L^2_{\frac{3}{2}}} ||F_A||_{L^2} + ||\nabla\phi_h||_{C^0} ||F_A||_{L^2} \Big). \end{aligned} \tag{35}$$

By Corollary 2.5.2 this implies that

(36) $$\begin{aligned} \|\alpha\|_{L^2_1} \leq & \frac{5}{4}C\Big(\|a\|_{L^2_{\frac{3}{2}}}\|\alpha\|_{L^2_1}\| + \|\phi_h - Id\|_{C^0}\|\nabla_\Gamma f_A\|_{L^2} \\ & + \|a\|_{L^2_{\frac{3}{2}}}\|\nabla_\Gamma f_A\|_{L^2} + \|\nabla\phi_h\|_{C^0}\|\nabla_\Gamma f_A\|_{L^2}\Big). \end{aligned}$$

Now supposing that U_Γ was chosen so that

$$C max_{\{A\in U_\Gamma\}}\|A-\Gamma\|_{L^2_{\frac{3}{2}}} < (1/4),$$

Inequality 36 implies the existence of a constant $C' > 0$ (depending only on g_0) so that

(37) $$\|\alpha(t)\|_{L^2_1} \leq C'(\|h\|_{C^0} + \|a\|_{L^2_{\frac{3}{2}}}(1+\|h\|_{C^1}))\|\nabla_A f_\Gamma\|_{L^2}$$

Clearly, Inequality 30 follows provided that we have chosen U_Γ and b sufficiently small.

Next we prove Inequality 32.

$$\begin{aligned} & d^*_\Gamma d_\Gamma(\alpha - \Theta(A)) + d^*_\Gamma([a, \alpha - \Theta(A)]) = \\ & \quad -*\bigwedge((Id\otimes(*\phi_h * -Id))(\nabla_\Gamma F_A) + (a\otimes F_A)) - *[\nabla\phi_h, (F_A)] - *[a, F_A] \\ & = \ *\bigwedge((Id\otimes(*\phi_h * -Id))(\nabla_\Gamma F_A) - *[\nabla\phi_h, (F_A)]). \end{aligned}$$

Hence as above there is a constant C so that for all sufficiently small gradient flow patches U_Γ and for $i = 1, 2$ we have the estimates

(38) $$\|\alpha - \Theta(A)\|_{L^2_i} \leq C\|h(t)\|_{C^1}\|F_{A(t)}\|_{L^2_{i-1}}.$$

This inequality with $i = 2$ and the embedding $L^2_2 \hookrightarrow C^0$ on N and Inequality 17 implies Inequality 32.

Next we prove Inequality 31. Let $\wp(t)$ be defined as the error term as in Equation 23. Clearly, $\wp(t) \in Ker(\Delta_\Gamma)^\perp$. From Equations 5 and 9 we have

(39) $$\wp(t) = *F_{A(t)} - \phi_h(*F_{A(t)}) + d_{A(t)}\alpha(t) - d_{A(t)}\Theta(A(t))$$

Let us prove that $\wp(t) \in L^2(N)$ for all $t \in [T_0, T_1)$. Since $A \in L^2_{(1,1)}([T_0, T_1) \times N)$ we have that $F_A \in L^2_1([T_0, T_1) \times N)$ while $\frac{\partial A}{\partial t} \in L^2_{(1,0)}([T_0, T_1) \times N)$ and $d_A(\Theta(A)) \in L^2_{(2,1)}([T_0, T_1) \times N)$ so $\wp$ is in the smallest of these, namely $L^2_{(1,0)}([T_0, T_1) \times N)$.

By Equation 39 and Inequality 38 with $i = 1$ we have

$$\begin{aligned} ||\wp(t)||_{L^2} &\leq C_1(||h||_{C^0}||F_{A(t)}||_{L^2} \\ &\quad + C_2(1 + ||a(t)||_{L^2_{\frac{3}{2}}})||\alpha(t) - \Theta(A(t))||_{L^2_1} \\ &\leq C_1(||h||_{C^0}||F_{A(t)}||_{L^2} + C_2C(1 + ||a||_{L^2_{\frac{3}{2}}})||F_{A(t)}||_{L^2} \end{aligned}$$

Combining this with Corollary 17 proves Inequality 31.

□

3.3. Regularity for ASD connections in standard form

In the section we prove that the standard local elliptic estimates hold for ASD connections in standard form. In particular we will show how L^2_k-estimates on small bands can be derived from L^2-estimates on larger bands. One step in the proof of regularity is to estimate the dt-component of a g-ASD connection in standard form in terms of the path of connections along the slices $\{t\} \times N$. Using the same circle of ideas as in the proof of Lemma 3.2.1 we prove the following lemma which accomplishes this. The calculations actually establish a more general result which we will need in Section V, so we include it here.

LEMMA 3.3.1. *For any C^∞ flat connection Γ on η there are constants $C, b > 0$ and a coordinate patch U_Γ centered at Γ so that the following holds. Suppose that $g = g_0 + h$ is a C^3 Riemannian metric with $||h||_{C^3} \leq b$ and that $\tilde{A}_i(t) + \Theta(\tilde{A}_i(t))dt$, for $i = 1, 2$, is a pair of g_0 ASD connections in in standard form with respect to U_Γ while $A_i(t) + \alpha_i(t)dt$, for $i = 1, 2$, are g ASD connections in standard form with* *respect to U_Γ. Then for any $0 \leq k \leq 3$ we have*

$$(40) \quad \|\alpha_1(t) - \Theta(\tilde{A}_1(t))\|_{L^2_k} \leq$$
$$C(\|A_1(t) - \tilde{A}_1(t)\|_{L^2_k} + \|F_{A_1(t)}\|_{L^2_{k-1}} + \|A_1(t) - \tilde{A}_1(t)\|_{L^2_k}\|F_{A_1(t)}\|_{L^2_{k-1}})$$

and

$$\begin{aligned}
&\|\frac{\partial^k}{\partial t^k}\Big((\alpha_1(t) - \Theta(\tilde{A}_1(t))) - (\alpha_2(t) - \Theta(\tilde{A}_2(t)))\Big)\|_{L^2} \\
&\leq C(\|\frac{\partial^k}{\partial t^k}\Big((A_1(t) - \tilde{A}_1(t)) - (A_2(t) - \tilde{A}_2(t))\Big)\|_{L^2} \\
&+\|h\|_{C^{k-1}}\|F_{A_1(t)} - F_{A_2(t)}\|_{L^2})
\end{aligned}$$

in the sense that if the right-hand-side of either inequality is finite then so is the left-hand-side and the inequality holds.

Proof. As usual we write $A_1 = \Gamma + a_1$ and $\tilde{A}_1 = \Gamma + \tilde{a}_1$. Also let $\alpha_1 = \Theta(A_1)$ and $\tilde{\alpha}_1 = \Theta(\tilde{A}_1)$. Using Equation 33 we see that the following equation holds.

$$\begin{aligned}
d^*_\Gamma d_\Gamma(\alpha_1 - \tilde{\alpha}_1) &= d^*_\Gamma([a_1, \alpha_1] - [\tilde{a}_1, \tilde{\alpha}_1]) \\
&\quad + * \bigwedge ((Id \otimes (*\phi_h * - Id))(\nabla_\Gamma F_{A_1}) + a_1 \otimes F_{A_1})) \\
&\quad - * [\nabla \phi_h, (F_{A_1})] - *(\tilde{a}_1 \wedge *F_{\tilde{A}_1}) \\
&= d^*_\Gamma([a_1 - \tilde{a}_1, \alpha_1] - [\tilde{a}_1, \tilde{\alpha}_1 - \alpha_1]) \\
&\quad + * \bigwedge ((Id \otimes (*\phi_h * - Id))(\nabla_\Gamma F_{A_1})) \\
&\quad - * [\nabla \phi_h, (F_{A_1})] + *((a_1 - \tilde{a}_1) \wedge F_{A_1}) - *(\tilde{a}_1 \wedge (F_{\tilde{A}_1} - F_{A_1})).
\end{aligned}$$

The Inequality 40 now follows from the Sobolev multiplication theorems. The other inequality follows from an analogous argument. □

Now we are in position to prove a version of the standard bootstrap lemma for ASD connections in standard form. See for example [29] or [10] for similar results for connections in Coulomb gauge. These results provide a convenient way of passing from L^2-estimates to L^2_k- estimates and Hölder estimates for flow lines in a slice.

LEMMA 3.3.2. *Fix $0 < \alpha < 1$. Let Γ be a C^∞ flat connection on η. For all sufficiently small gradient flow patches U_Γ there are constants $\kappa, K, b > 0$ so that the following holds. Let $g = g_0 + h$ be a C^3 Riemannian metric with $||h||_{C^3} < b$ and let $\omega_i = A_i(t) + \alpha_i(t)dt$, for $i = 1, 2$, be g-ASD connections on $[T-1, T+1] \times \eta$ in standard form with respect to U_Γ. Suppose that there are flat connections Γ_i in U_Γ for $i = 1, 2$ satisfying:*

(1) $\int_{T-1}^{T+1} \int_N (|F_{\omega_i}|_g)^2 \leq \kappa$ *and*

(2) $||\omega_i - \Gamma_i||_{L^2([T-1,T+1]\times N)} \leq \kappa$.

Then the following estimates hold;

$$||\omega_1 - \omega_2||_{L^2_3([T-\frac{1}{2},T+\frac{1}{2}]\times N)} \leq K||\omega_1 - \omega_2||_{L^2([T-1,T+1]\times N)}.$$

$$||\omega_1 - \omega_2||_{C^{(3,\alpha)}([T-\frac{1}{2},T+\frac{1}{2}]\times N)} \leq K||\omega_1 - \omega_2||_{L^2([T-1,T+1]\times N)}.$$

Before we prove Lemma 3.3.2 a few preliminary remarks are in order. For $s \in (0, 1]$ let $M_s = [T-s, T+s] \times N$ Over M_1 consider the product bundle $P = [T-1, T+1] \times \eta$ and the corresponding ASD deformation operator.

$$D = (P(g)_+ d_\Gamma, d^*_\Gamma) \colon \Omega^1(adP) \to \Omega^2_+(adP) \oplus \Omega^0(adP).$$

D is an elliptic operator (with C^3 coefficients with our choice of g since d^* is taken with respect to the smooth product metric g_0) and hence we have the Friedrichs Lemma [16].

LEMMA 3.3.3. *Fix $0 \leq k \leq 3$, $1 \geq s_1 > s_2 > 0$, and $0 < \alpha < 1$. Then there is a constant $C > 0$ so that for all $u \in L^2_k(M_{s_1})$ we have*

$$||u||_{L^2_k(M_{s_2})} \leq C(||Du||_{L^2_{k-1}(M_{s_1})} + ||u||_{L^2(M_{s_1})}),$$

and

$$||u||_{C^{(k,\alpha)}(M_{s_2})} \leq C(||Du||_{C^{(k-1,\alpha)}(M_{s_1})} + ||u||_{C^{(0,\alpha)}(M_{s_1})})$$

LEMMA 3.3.4. *Fix $s_1 > s_2$. Let Γ be a C^∞ flat connection on η. For any sufficiently small coordinate patch U_Γ there are constants $K_2, \kappa_2, b > 0$ so that the following holds. Let $||g - g_0||_{C^3} < b$ and for $i = 1, 2$ let $\omega_i = A_i(t) + \alpha_i(t)dt$ be g-ASD connections in standard form with respect to U_Γ satisfying:*

$$||\omega_i - \Gamma||_{L^4(M_{s_1})} \leq \kappa_2$$

for $i = 1, 2$. Then

$$||\omega_1 - \omega_2||_{L^2_3(M_{s_1})} \leq K_2||\omega_1 - \omega_2||_{L^2(M_{s_2})},$$

and

$$||\omega_1 - \omega_2||_{C^{(3,\alpha)}(M_{s_1})} \leq K_2||\omega_1 - \omega_2||_{L^2(M_{s_2})}.$$

Proof. Let $\beta\colon M \to [0,1]$ be a C^∞ cutoff function with $\beta(M_{s_2}) = 1$ and $\beta = 0$ on a neighborhood of ∂M_{s_1}. Since the ω_i are g-ASD we have $P_+ d_\Gamma \omega_i = -P_+((\omega_i - \Gamma) \wedge (\omega_i - \Gamma))$. Also since the ω_i are in standard form we have $d_\Gamma^* \omega_i = -\frac{\partial \alpha_i}{\partial t}$. Combining these observations with the Friedrichs lemma and Inequality 3.3.1 implies the following for some constant $C > 0$.

$$\begin{aligned}
&||\beta(\omega_1 - \omega_2)||_{L^2_1(M_{s_1})} \\
&\qquad \leq C(||D(\beta(\omega_1 - \omega_2))||_{L^2(M_{s_1})} + ||\beta(\omega_1 - \omega_2)||_{L^2(M_{s_1})}) \\
&\qquad \leq C(||\sigma(D, d\beta)(\omega_1 - \omega_2)||_{L^2(M_{s_1})} + ||\beta D(\omega_1 - \omega_2))||_{L^2(M_{s_1})} \\
&\qquad\quad + ||\beta(\omega_1 - \omega_2)||_{L^2(M_{s_1})}) \\
&\qquad \leq C(||\sigma(D, d\beta)(\omega_1 - \omega_2)||_{L^2(M_{s_1})} \\
&\qquad\quad + ||\beta(\omega_1 \wedge \omega_1 - \omega_2 \wedge \omega_2)||_{L^2(M_{s_1})} + ||\beta \frac{\partial}{\partial t}(\alpha_1 - \alpha_2)||_{L^2(M_{s_1})} \\
&\qquad\quad + ||\beta(\omega_1 - \omega_2)||_{L^2(M_{s_1})} \leq
\end{aligned}$$

$$\begin{aligned}
&\leq C_1(\|\sigma(D,d\beta)(\omega_1-\omega_2)\|_{L^2(M_{s_1})} \\
&\quad + (\|\omega_1-\Gamma\|_{L^4(M)}+\|\omega_2-\Gamma\|_{L^4(M_{s_1})})\|\beta(\omega_1-\omega_2)\|_{L^4(M_{s_1})} \\
&\quad + \epsilon\|\beta\frac{\partial}{\partial t}(A_1(t)-A_2(t))\|_{L^2(M_{s_1})} \\
&\quad + \|h\|_{C^3}\|F_{\omega_1}-F_{\omega_2}\|_{L^2(M_{s_1})}\|\beta(\omega_1-\omega_2)\|_{L^2(M_{s_1})}) \\
&\leq C_2(\|\sigma(D,d\beta)(\omega_1-\omega_2)\|_{L^2(M_{s_1})} \\
&\quad + (2\kappa_2+\epsilon+C_3\|h\|_{C^3})\|\beta(\omega_1-\omega_2)\|_{L^2_1(M_{s_1})} \\
&\quad + \|\beta(\omega_1-\omega_2)\|_{L^2(M_{s_1})}).
\end{aligned}$$

Here $\sigma(D,d\beta)$ denotes the symbol of D evaluated on the cotangent vector $d\beta$, and $\epsilon=\epsilon(b)$ from Lemma 3.3.1. Thus if we choose K_2,κ_2,b sufficiently small so that $C_2(2\kappa_2+\frac{1}{8}+C_3\|h\|_{C^3})\leq\frac{1}{2}$, then we have

$$\begin{aligned}
\|\omega_1-\omega_2\|_{L^2_1(M_{s_2})} &\leq \|\beta(\omega_1-\omega_2)\|_{L^2_1(M_{s_1})} \\
&\leq 2C_2(\|\sigma(D,d\beta)(\omega_1-\omega_2)\|_{L^2(M_{s_1})}+\|\beta(\omega_1-\omega_2)\|_{L^2(M_{s_1})}) \\
&\leq C_4\|\omega_1-\omega_2\|_{L^2(M_{s_1})}.
\end{aligned}$$

Arguing similarly we may derive the following estimate for a possibly smaller choice of K_2,κ_2,b

$$\|\omega_1-\omega_2\|_{L^2_3(M_{s_2})}\leq K_2\|\omega_1-\omega_2\|_{L^2(M_{s_1})}.$$

The proof of the Hölder estimate is the same as above given the embedding $L^2_3\hookrightarrow C^{(0,\alpha)}$ for $0<\alpha<1$. □

We are now ready to prove Lemma 3.3.2

Proof. The choice of the coordinate patch U_Γ is determined by the following conditions. First, Lemma 3.3.4 is to hold for U_Γ, $s_1=\frac{3}{4}$ and $s_2=\frac{1}{2}$ with the constants K_2,κ_2. Second, for all flat connections $\Gamma'\in U_\Gamma$ we have $\|\Gamma'-\Gamma\|_{L^4(M_{\frac{3}{4}})}<\kappa_2/3$. That such a choice is possible follows from the equivalence of the L^4- topology and the standard topology on flat connections.

Next find $\delta > 0$ so that for all flat connections $\Gamma', \Gamma'' \in U_\Gamma$ we have that if $||\Gamma' - \Gamma''||_{L^2(M_{\frac{3}{4}})} \leq \delta$ then $||\Gamma' - \Gamma''||_{L^4(M_{\frac{3}{4}})} \leq \kappa_2/3$. The constant κ is determined by the following conditions. First, $\kappa \leq \delta/2$. Second, we require that if ω is a g-ASD connection on $[T-1, T+1] \times \eta$ with $\int_{M_1} |F_\omega|_g^2 \leq \kappa$ then there is a flat connection Γ' so that $||\omega - \Gamma'||_{L^2(M_{\frac{3}{4}})} \leq \delta/2$ and $||\omega - \Gamma'||_{L^4(M_{\frac{3}{4}})} \leq \kappa_2/3$. Uhlenbeck's compactness theorem guarantees that this last condition on κ can be satisfied.

Let ω_1, ω_2 and Γ_1 and Γ_2 satisfy the hypothesis Lemma 3.3.2. By the above discussion we have flat connections Γ_i', $i = 1, 2$ such that

$$||\omega_i - \Gamma_i'||_{L^2(M_{\frac{3}{4}})} \leq \delta/2$$
$$||\omega_i - \Gamma_i'||_{L^4(M_{\frac{3}{4}})} \leq \kappa_2/3$$

Thus, we have

$$\begin{aligned} ||\Gamma_i - \Gamma_i'||_{L^2(M_{\frac{3}{4}})} &\leq ||\omega_i - \Gamma_i||_{L^2(M_{\frac{3}{4}})} + ||\omega_i - \Gamma_i'||_{L^2(M_{\frac{3}{4}})} \\ &\leq \delta/2 + \delta/2 = \delta. \end{aligned}$$

Hence, by our choice of U_Γ and δ we have

$$||\Gamma_i - \Gamma_i'||_{L^4(M_{\frac{3}{4}})} \leq \kappa_2/3.$$

This yields,

$$\begin{aligned} ||\omega_i - \Gamma||_{L^4(M_{\frac{3}{4}})} &\leq ||\omega_i - \Gamma_i'||_{L^4(M_{\frac{3}{4}})} + ||\Gamma_i - \Gamma_i'||_{L^4(M_{\frac{3}{4}})} + ||\Gamma - \Gamma_i||_{L^4(M_{\frac{3}{4}})} \\ &\leq \kappa_2/3 + \kappa_2/3 + \kappa_2/3 = \kappa_2 \end{aligned}$$

Thus by the choice of U_Γ and κ_2 Lemma 3.3.4 applies to the pair $M_{\frac{1}{2}}$ and $M_{\frac{3}{4}}$ to give the desired result. The proof of the Hölder estimate is the same as above once we have the embedding $L_3^2 \hookrightarrow C^{(0,\alpha)}$. □

3.4. Completion of the proof

To begin choose the coordinate patch U_Γ and the constant b so small that Lemmas 3.2.1 and 3.3.2 hold. We also require that $C_0 b \leq \frac{1}{2}$ where C_0 is the constant of Lemma 3.2.1. Let B denote the space of C^3 riemannian metrics $g = g_0 + h$ so that $||h||_{C^3} \leq b$ Further restrict U_Γ so that for all ω in standard form with respect to U_Γ we have $||\omega - \Gamma||_{L^4([T-1,T+1)\times N)} < \kappa_2$ where κ_2 is the constant of Lemma 3.3.4 for $s_1 = \frac{1}{2}$ and $s_2 = 1$.

Fix $g = g_0 + h$ a C^3 Riemannian metric with $||h||_{C^3} < b$ Let ω be an L^2_2 g-ASD connection on $[T_0 - 1, T_1 + 1) \times \eta$ satisfying the hypothesis of Theorem 2.6.3.

Since Lemma 3.2.1 holds there is a gauge transformation $\sigma \in \mathcal{G}_{(1,2)}([T_0 - 1, T_1 + 1) \times \eta)$ so that $\sigma^*\omega = A + \alpha dt$ is in standard form with respect to U_Γ. Now given our choice of U_Γ and g Lemma 3.3.2 applied to any band $[T - \frac{1}{2}, T + \frac{1}{2}] \times N$ implies that $\sigma^*\omega$ is $C^{(3,\alpha)}$. It follows that σ is in fact an $L^2_{3,loc}$ gauge transformation. The remaining estimates following immediately from those in Lemma 3.2.1.

3.5. Related estimates

In this last section we record two estimates which are related to the ones above and which will be of use later. The first is relates the norm of the curvature on a slice to the energy on a band of fixed size centered at the slice.

LEMMA 3.5.1. *There are constants $\epsilon_1, C_1 > 0$ such that for any $g \in B$ and any $T \geq 1$ and any g- ASD connection ω on $E|([T-1, T+1) \times N)$ with*

$$\int_{T-1}^{T+1} \int_N (|F_\omega|_g)^2 < \epsilon_1$$

then

$$\left(||F_{\omega|\{T\}\times N}||_{L^2_1}\right)^2 \leq C_1^2 \int_{T-1}^{T+1} \int_N (|F_\omega|_g)^2$$

where $|F_\omega|_g$ is the pointwise norm with respect to g.

Proof. This follows immediately from Proposition 8.3 in [12] and the fact that for any $g \in B$ the g-norm differs from the g_0-norm by at most a fixed multiplicative constant, independent of g. □

We also have a more complicated version of Lemma 3.3.2. It is proved in the same way.

LEMMA 3.5.2. *Fix Γ a flat connection on η. Then for all sufficiently small patches U_Γ there are constants $\kappa, K > 0$ so that the following holds. Let $g \in B$ and let ω_i for $i = 1, 2$ be g-ASD connections on $[T-1, T+1) \times \eta$ in standard form with respect to U_Γ and $\hat{\omega}_i$ for $i = 1, 2$ be g_0-ASD connections. Suppose there are flat connections Γ_i for $i = 1, \ldots, 4$ in U_Γ so that the following hold.*

(1) $\int_{M_1} (|F_{\omega_i}|_g)^2 \leq \kappa$ *for* $i = 1, \ldots, 4$.

(2) $||\omega_i - \Gamma_i||_{L^2(M_1)} \leq \kappa$ *for* $i = 1, \ldots, 4$.

Then the following estimate holds:

$$\begin{aligned} ||(\omega_1 - \omega_2) - (\omega_3 - \omega_4)||_{L^2_3(M_{\frac{1}{2}})} \quad &\leq \quad K||(\omega_1 - \hat{\omega}_1) - (\omega_2 - \hat{\omega}_2)||_{L^2(M_1)} \\ &\quad + K||h||_{C^3}(||F_{\omega_1}||_{L^2(M_1)} + ||F_{\omega_2}||_{L^2(M_1)}) \end{aligned}$$

CHAPTER 4

Existence of an asymptotic value at infinity

Chapter 2 introduced the basic material relating the ASD equation for connections on $[0,\infty)\times\eta$ to a gradient flow equation for the restriction f_Γ of the Chern-Simon's function to the slice U_Γ through a flat connection Γ in the space of connections $\mathcal{A}(\eta)$. In this chapter we use this analysis to prove the existence of an asymptotic value at infinity for every finite energy ASD connection. Let B be a neighborhood of g in the space of C^3-metrics as in Theorem 2.6.3. We call any metric $g\in B$ a *cylindrical end metric on X*.

We denote by $\chi(N)\subset\mathcal{B}(N)$ the subspace of gauge equivalence classes of flat connections. We call it *the character variety* for N. Analogously, we have the space of gauge equivalence classes of based flat connections $\mathcal{R}(N)\subset\mathcal{B}^0(N)$. It is called the *representation variety* of N.

Here is the main theorem that we shall prove in this chapter.

THEOREM 4.0.1. *Let $g\in B$. Let ω be a $L^2_{2,loc}$ g-ASD connection of finite energy on the principal G-bundle $E\to X$. Recall that $w\in H^2(X;\pi_1(G))$ is the associated characteristic class. For any $L^2_{3,loc}$ bundle isomorphism $\varphi\colon[0,\infty)\times\eta\cong E|([0,\infty)\times N)$, the associated path $[\varphi^*\omega|\{t\}\times\eta]$ in $\mathcal{B}(N)$ has a limit as $t\to\infty$ which is the gauge equivalence class of a flat connection. Associating this limit to*

$[\omega]$ defines a continuous map

$$\partial\colon \mathcal{M}_w(X,g) \to \chi(N).$$

The proof breaks into two basic steps. We first appeal to the compactness theorem of Uhlenbeck to deduce that the path $[\varphi^*\omega|\{t\} \times \eta]$ of connections on η spirals in towards and limits on a component of $\chi(N)$. The import of this is that we can work for a short time (*a priori*) in a small ball in a slice centered at a flat connection Γ. At this point we appeal to analogues of L. Simon's work on the asymptotics of gradient flows (see [25]). His results imply that the path stays inside of the slice at Γ, and converges to a unique flat connection in that slice.

4.1. Weak convergence to the character variety

The method of proof of Theorem 4.0.1 is to appeal to analogues of results of L. Simon about asymptotics of gradient flows and perturbations of gradient flows. Before we can apply these analogues we need a preliminary result based on Uhlenbeck compactness concerning weak asymptotic properties of the path $[\varphi^*\omega|\{t\} \times \eta] \in \mathcal{B}(N)$. The purpose of this section is to prove the needed weak convergence result.

LEMMA 4.1.1. *Let Γ be a C^∞ flat connection on η, and let U_Γ be a coordinate patch centered at Γ. Let $V_\Gamma \subset U_\Gamma$ be an open subset whose closure in S_Γ is contained in U_Γ. Then for each $\delta_0 > 0$ there is $\delta_1 > 0$ such that for any connection $A \in V_\Gamma$ with $||F(A)||_{L^2_{\frac{1}{2}}} < \delta_1$ there is a C^∞ flat connection $\Gamma' \in U_\Gamma$ with*

$$||A - \Gamma'||_{L^2_{\frac{3}{2}}} < \delta_0.$$

Proof. Let $A_n \in V_\Gamma$ be a sequence with $lim_{n\to\infty}||F(A)||_{L^2_{\frac{1}{2}}} = 0$. We shall establish the lemma by showing that for every such sequence there is a subsequence converging to flat connection in U_Γ. First of all, by Uhlenbeck's weak compactness and the compactness of the character variety, there is a subsequence of the A_n

converging weakly in U_Γ to a flat connection Γ'. We write $A_n = \Gamma' + a_n$. We know that $||a_n||_{L^2_{1+\alpha}} \mapsto 0$ as $n \mapsto \infty$ for any $\alpha < 1/2$. We also know

$$d^*_\Gamma a_n = 0$$

$$d_{\Gamma'}(a_n) = F(A_n) - \frac{1}{2} a_n \wedge a_n.$$

That $||a_n||_{L^2_{\frac{3}{2}}} \mapsto 0$ now follows by the Sobolev multiplication theorem and standard elliptic arguments. □

LEMMA 4.1.2. *Let $g \in B$. We let ω be a finite energy, g-ASD $L^2_{2,loc}$-connection on the principal G-bundle $[0, \infty) \times \eta$. We write*

$$\omega = A(t) + \alpha(t)dt,$$

where $A(t)$ is a path in $\mathcal{A}_{\frac{3}{2}}(\eta)$. Let Γ be a C^∞ flat connection on η. Let $CS = CS_{\Gamma_0}$ be the Chern-Simons function with respect to some fixed connection Γ_0. Then the following hold:

1. *$CS(A(t))$ is a non-increasing function of t and $lim_{t \mapsto \infty} CS(A(t))$ exists.*
2. *The path $[A(t)]$ in $\tilde{\mathcal{B}}(N)$ has non-empty ∞-limit set contained in a component $\tilde{\chi}_0(N)$ of the pre-image $\tilde{\chi}(N)$ of the character variety $\chi(N) \subset \mathcal{B}(N)$. (By definition the ∞-limit set is the set of all limits of sequences $[A(t_n)]$ where $t_n \mapsto \infty$ as $n \mapsto \infty$.)*
3. *The limiting value of $CS(A(t))$ is equal to the value of CS on $\tilde{\chi}_0(N)$.*

Proof. First, notice that since ω is g-ASD for any $0 \leq t_0 \leq t_1 \leq \infty$ we have

$$\begin{aligned} CS(\omega|\{t_0\} \times N) - CS(\omega|\{t_1\} \times N) &= \int_{t_0}^{t_1} \int_N -tr(F_\omega \wedge *_g F_\omega) \\ &= ||F_\omega|([t_0, t_1] \times N)||^2_{L^2,g} \end{aligned} \tag{41}$$

where $|| \, ||_{L^2,g}$ means the L^2-norm with respect to g. Part 1 is immediate from this and the fact that ω has finite energy.

Fix a neighborhood $\mathcal{N}$ of $\chi(N)$ in $\mathcal{B}(N)$. By Lemmas 4.1.1 and 3.5.1, and the compactness of $\chi(N)$, there is $T > 0$ such that $[A(t)] \in \mathcal{N}$ for all $t \geq T$. This proves that the ∞-limit set for the curve $[A] \in \mathcal{B}(N)$ is contained in a component of $\chi(N)$. Since $\chi(N)$ is compact, it follows that this ∞-limit set must be non-empty. Part 3 is immediate from Parts 1 and 2. □

One consequence of the weak convergence to the character variety is that it lets us work with connections of a fixed energy. Recall that the energy of a connection ω is the integral

$$-\int_X trF_\omega \wedge *F_\omega$$

where the trace is taken in the adjoint representation of the Lie algebra of G. In the case when ω is ASD its energy is equal to $-8\pi^2 p_1(ad\omega)$.

We fix a real number ℓ. We define $\mathcal{M}_{\ell,w}(E,g)$ to be the subspace of $\mathcal{M}_w(E,g)$ of connections with energy $2\pi^2\ell$. (If X is a closed manifold, $G = SU(2)$ then $\ell = c_2(E)$.) Fix a connection ω_0 on E which is isomorphic to a product connection Γ_0 in the cylindrical end of X. Notice that $p_1(ad\omega_0)$ is a compactly supported form on X.

COROLLARY 4.1.3. *$\mathcal{M}_{\ell,w}(E,g)$ is non-empty only if $-4\ell - \int_X p_1(ad\omega_0)$ is congruent modulo Z to the value of CS on a flat connection on η. Furthermore, $\mathcal{M}_{\ell,w}(E,g)$ is a union of components of $\mathcal{M}_w(E,g)$.*

Proof. The first sentence is clear from Lemma 4.1.2. The second follows since we have required that the energy be a continuous function on $\mathcal{M}_w(E,g)$. □

4.2. L. Simon's theorems on asymptotics

Theorem 4.0.1 is a strengthening of the weak limiting result, Lemma 4.1.2, of the last section. It is based upon an infinite dimensional version of the Lojaszewicz inequality, essentially due to L. Simon. The version of Simon's result that we use is the following.

PROPOSITION 4.2.1. *Let* Γ *be a* C^∞*-flat connection on* η*, and let* U_Γ *be a coordinate patch centered at* Γ*. Then there is a neighborhood* $V_\Gamma \subset U_\Gamma$ *of* Γ *and constants* $0 < \beta \leq 1$ *and* $0 < \theta \leq 1/2$ *so that if* $A \in V_\Gamma$ *then there is a* C^∞ *flat connection* $\Gamma' \in U_\Gamma$ *so that:*

$$||A - \Gamma'||_{L^2} \leq (||\nabla_A f_\Gamma||_{L^2})^\beta \tag{42}$$

$$|f_\Gamma(A)|^{(1-\theta)} \leq ||\nabla_A f_\Gamma||_{L^2}. \tag{43}$$

Proof. We claim that Simon's proof of the Łojaszewicz inequality, Lemma 2 in [25], adapts to our situation to give the two inequalities of this proposition. More explicitly, Simon proves the following.

PROPOSITION 4.2.2. *Let* $E \to M$ *be a hermitian vector bundle over a compact manifold. Let* $f\colon C^\infty(M; E) \to R$ *be a continuous function which extends to some Sobolev completion* $H = L^2_k(M; E)$ *and is real analytic on some open subset* $V \subset H$*. Furthermore, suppose that the formal adjoint of the Hessian of* f *at each* $s \in V$ *is elliptic. Let* $C \subset H$ *be the critical set of* f*. Then for each* $c \in C \cap V$ *there is a neighborhood* $U_c \subset V$ *of* c *and constants* $0 < \beta \leq 1$ *and* $0 < \theta \leq 1/2$ *such that for every* $A \in U_c$*:*

$$inf_{\{c' \in C \cap U_c\}} ||A - c'||_{L^2} \leq (||\nabla_A f||_{L^2})^\beta$$

and

$$|f(A) - f(c)|^{(1-\theta)} \leq ||\nabla_A f||_{L^2}$$

The only difference in our context is that we are working on an open subset U_Γ in the slice S_Γ rather than on an entire space of sections. By the Appendix to this chapter we know that f_Γ is a real analytic function on $U_\Gamma \subset S_\Gamma$ and that the formal adjoint of its Hessian is elliptic on the tangent space $Ker(d^*_\Gamma) \subset \Omega^1(N)$ to U_Γ. Simon's proof adapts *mutatis-mutandis* to this context. This proves that for a sufficiently small neighborhood $V_\Gamma \subset U_\Gamma$ of Γ there are constants $0 < \beta < 1$ and $0 < \theta \leq 1/2$ such that Inequalities 42 and 43 hold where the gradient is computed with respect to the standard L^2-inner product. We however are working with a continuously varying inner product $\langle , \rangle_{\{\Gamma, A\}}$ on the tangent space at A. The important fact is that for $A \in V_\Gamma$ this inner product is uniformly commensurate with the standard L^2-inner product, and thus Inequalities 42 and 43 hold up to a multiplicative constant. Since $\nabla_\Gamma f_\Gamma = 0$, by choosing V_Γ sufficiently small, we can arrange that $||\nabla_A f_\Gamma||$ is arbitrarily small for all $A \in V_\Gamma$. Once we have done this we can absorb any multiplicative constant by decreasing β and θ. This completes the proof of Proposition 4.2.1. □

We have the following gauge invariant form of Proposition 4.2.1.

COROLLARY 4.2.3. *Let Γ be a C^∞-flat connection on η, and let U_Γ be a coordinate patch centered at Γ. Then there is a neighborhood $V_\Gamma \subset U_\Gamma$ of Γ and constants $0 < \beta \leq 1$ and $0 < \theta \leq 1/2$ so that if $A \in V_\Gamma$ then there is a C^∞ flat connection $\Gamma' \in U_\Gamma$ so that:*

$$||A - \Gamma'||_{L^2} \leq (||F_A||_{L^2})^\beta \tag{44}$$

$$|f_\Gamma(A)|^{(1-\theta)} \leq ||F_A||_{L^2} \tag{45}$$

Proof. This is immediate from Proposition 4.2.1 and Corollary 2.5.2 and the argument, given above, which allows one to absorb multiplicative constants by choosing U_Γ sufficiently small and lowering θ and β. □

This gauge invariant form of Inequality 42 is an analogue of the fundamental lemma of Uhlenbeck's [29]. In Uhlenbeck's case the trivial connection is a non-degenerate minimum for the Yang Mills functional on the ball, and consequently one can take $\beta = 1$. In our case, the flat connections are possibly degenerate minima of the Yang Mills functional for η and hence in Estimate 42 one can not in general take $\beta = 1$ even at the expense of allowing a multiplicative constant. We work with the non-gauge invariant form of the proposition because this fits better with Simon's results.

These estimates lead to the following analogue of Simon's theorem on finite length.

THEOREM 4.2.4. *Let Γ be a C^∞ flat connection on η, and let U_Γ be a coordinate patch centered at Γ. Then there is a neighborhood $V_\Gamma \subset U_\Gamma$ of Γ and a constant $\theta = \theta(\Gamma)$ with $0 < \theta \leq 1/2$ such that the following holds. If $A = A(t)$ for $T_0 \leq t < T_1$ is a C^1-path in V_Γ solving the perturbed gradient flow equation, Equation 23, with the error estimate given by Equation 26, then:*

(1) *If $CS(A(T_0)) > CS(\Gamma) > CS(A(T_1))$ then*

$$\int_{T_0}^{T_1} \left\| \frac{\partial A(t)}{\partial t} \right\|_{L^2} dt \leq (2/\theta) \left\{ \Big(CS(A(T_0)) - CS(\Gamma) \Big)^\theta + \Big(CS(\Gamma) - CS(A(T_1)) \Big)^\theta \right\}. \tag{46}$$

(2) *Otherwise, we have*

$$\int_{T_0}^{T_1} \left\| \frac{\partial A(t)}{\partial t} \right\|_{L^2} dt \leq (2/\theta) \Big(\big| CS(A(T_0)) - CS(A(T_1)) \big| \Big)^\theta. \tag{47}$$

COROLLARY 4.2.5. *With notation and assumptions as in Theorem 4.2.4 we have*

$$\int_{T_0}^{T_1} \left\| \frac{\partial A(t)}{\partial t} \right\|_{L^2} dt \leq \frac{4}{\theta} \left| CS(A(T_0)) - CS(A(T_1)) \right|^\theta. \tag{48}$$

4.3. Strong convergence to the character variety

The estimates of the previous section imply a much stronger convergence theorem.

THEOREM 4.3.1. *Let $g \in B$. Let Γ be a C^∞ flat connection on η, and let U_Γ be a coordinate patch centered at Γ. Then there is $0 < \epsilon_0 < 1$, a neighborhood $V_\Gamma \subset U_\Gamma$ of Γ, and a constant $0 < \theta \leq 1/2$, and $C > 0$ such that the following holds. Fix*

$1 \leq T_0 \leq T_1 \leq \infty$. Suppose ω is a g-ASD connection on $E|\big([0, T_1+1) \times N\big)$ which is $L^2_{2,loc}$ and satisfies

$$\int_{T_0-1}^{T_1+1} \int_N (|F_\omega|_g)^2 < \epsilon_0, \tag{49}$$

and $\omega\,\big|\big(E|(\{T_0\} \times N)\big)$ is $L^2_{\frac{5}{2}}$-bundle isomorphic to a connection $A_0 \in V_\Gamma$. Then there is an $L^2_{3,loc}$-bundle isomorphism

$$\varphi\colon [T_0, T_1) \times \eta \to E|[T_0, T_1) \times N$$

such that writing $\varphi^*\omega = A(t) + \alpha(t)dt$ we have

(1) *$\varphi^*\omega$ is a C^3-connection in standard form with respect to U_Γ.*
(2) *$A(T_0) = A_0$.*
(3) *The length of the path $A(t)$, $T_0 \leq t < T_1$ is finite. In fact,*

$$\int_{T_0}^{T_1} \left\| \frac{\partial A(t)}{\partial t} \right\|_{L^2} dt \leq (4/\theta) \left(\int_{T_0}^{\infty} \int_N (|F_\omega|_g)^2 \right)^\theta. \tag{50}$$

If $T_1 = \infty$ then in addition the following hold.

(4) *$A(t)$ converges in U_Γ to a C^∞ flat connection $\Gamma' \in U_\Gamma$.*
(5) *$\alpha(t)$ is a path in $(Ker\Delta_\Gamma)^\perp \cap L^2_{\frac{7}{4}} \subset \Omega^0_{\frac{7}{4}}(N)$ whose $L^2_{\frac{7}{4}}$-norm converges to 0.*

The bundle isomorphism φ is unique up to composition with the pullback of an automorphism $\tau_0 \in Stab(\Gamma)$.

Proof. For any $\epsilon > 0$, let $\mathcal{U}(\epsilon) \subset U_\Gamma$ be the open set consisting of all connections of the form $\Gamma' + a'$ where Γ' is a flat connection in U_Γ with $||\Gamma' - \Gamma||_{L^2} < \epsilon$ and $a' \in Kerd^*_\Gamma$ with $||a'||_{L^2_{\frac{3}{2}}} < \epsilon$. Since the L^2-norm and the $L^2_{\frac{3}{2}}$-norm are commensurate on the subset of flat connections in U_Γ, the $\mathcal{U}(\epsilon)$ form a cofinal system of neighborhoods about Γ in U_Γ. Choose $\epsilon > 0$ sufficiently small such that the closure of $\mathcal{U}(\epsilon)$ in S_Γ is contained in U_Γ.

Choose $Stab(\Gamma)$-invariant neighborhoods $V_\Gamma \subset \overline{V}_\Gamma \subset \mathcal{U}(\epsilon) \subset \overline{\mathcal{U}}(\epsilon) \subset V'_\Gamma \subset \overline{V}'_\Gamma \subset U_\Gamma$ such that Theorem 2.6.3 holds for V'_Γ, such that Theorem 4.2.4 holds for (V'_Γ, U_Γ)

with constant $0 < \theta \leq 1/2$, and such that V_Γ is contained in the ball of radius $\epsilon/3$ about Γ in the $L^2_{\frac{3}{2}}$-norm. Choose $\epsilon_0 > 0$ such that:

(1) $\epsilon_0 \leq \epsilon_1$ of Lemma 3.5.1,

(2) Lemma 4.1.1 holds for (V'_Γ, U_Γ) with $\delta_1 = C_1\sqrt{\epsilon_0}$ and $\delta_0 = \epsilon/3$ where C_1 is the constant given in Lemma 3.5.1.

(3) $(4/\theta)\epsilon_0^\theta \leq \epsilon/3$, and

(4) if Γ and Γ' are flat connections with $CS(\Gamma) > CS(\Gamma')$ then

$$CS(\Gamma) - CS(\Gamma') > \epsilon_0. \tag{51}$$

Let ω be a g-ASD connection on $E|([T_0 - 1, T_1 + 1) \times N$ satisfying Inequality 49 with $\omega|\{T_0\} \times \eta$ gauge equivalent to $A_0 \in V_\Gamma$. Let $[T_0, T_1')$ be the maximal sub-interval of $[T_0, T_1)$ with the property that for each $t \in [T_0, T_1')$ the connection $\omega\,\big|(E|(\{t\} \times N))$ is $L^2_{\frac{5}{2}}$-bundle isomorphic to a connection in V'_Γ. Applying Lemma 2.6.3 to the interval $[T_0, T_1')$ we produce an $L^2_{3,loc}$-bundle isomorphism $\varphi\colon [T_0, T_1') \times \eta \to E|[T_0, T_1') \times N$ such that $\varphi^*\omega = A + \alpha dt$ is a C^3-connection in standard form with respect to V'_Γ. Furthermore, A and α satisfy the perturbed gradient flow equation, Equation 23, with error Estimate 26. By hypothesis $A(T_0) \in V_\Gamma$ is $L^2_{\frac{5}{2}}$-gauge equivalent to $A_0 \in U_\Gamma$. It follows that $A(T_0)$ and A_0 are in fact gauge equivalent by an element in $\sigma \in Stab\Gamma$. Replacing φ by $\varphi \circ (Id \times \sigma)$ replaces A by $\sigma^* A$ and allows us to arrange that $A(T_0) = A_0$.

The upshot of the above paragraph is that the path $A(t)$, $T_0 \leq t < T_1'$, satisfies the hypotheses of Theorem 4.2.4. Hence, by Corollary 4.2.5

$$\left(\int_{T_0}^{T_1'} ||F_\omega||^2\right)^\theta = (CS(A(T_0) - CS(A(T_1')))^\theta \geq \frac{\theta}{4}||A(T_1') - A(T_0)||_{L^2}.$$

Hence, by our choice of ϵ_0 we have

$$||A(T_1') - A(T_0)||_{L^2} < \epsilon/3. \tag{52}$$

Now let us show that $T_1' = T_1$. Suppose that $T_1' < T_1$. Since $\overline{\mathcal{U}}(\epsilon) \subset V'_\Gamma$, there is

$T_0 < T < T_1'$ for which $A(T) \in V_\Gamma' - \mathcal{U}(\epsilon)$. Since $A(T_0) \in V_\Gamma'$ we have

(53) $$||A_0 - \Gamma||_{L^2_{\frac{3}{2}}} < \epsilon/3.$$

We shall arrive at a contradiction by showing that $A(T) \in \mathcal{U}(\epsilon)$. This contradiction will prove that $T_1' = T_1$, and hence establish Items 1 and 2 of the theorem.

Clearly, in light of Inequalities 52 and 53, it suffices to show that there is a flat connection $\Gamma' \in U_\Gamma$ with

(54) $$||A(T) - \Gamma'||_{L^2_{\frac{3}{2}}} < \epsilon/3.$$

Since $A(T) \in V_\Gamma'$ and Lemma 4.1.1 applies to V_Γ', to establish Inequality 54 it suffices to show that $||F(A)||_{L^2_{\frac{1}{2}}} < C_1\sqrt{\epsilon_0}$. But this is immediate from Lemma 3.5.1 and the fact that

$$\int_{T-1}^{T+1}\int_N (|F_\omega|_g)^2 \leq \int_{T_0-1}^{T_1+1}\int_N (|F_\omega|_g)^2 < \epsilon_0.$$

From the fact that ω is in standard form on $[T_0, T_1) \times N$ and Inequality 26 of Theorem 2.6.3, and Equation 41, we see that $A(t), T_0 \leq t < T_1$ satisfies the hypotheses of Theorem 4.2.4. Item 3 of the theorem is immediate from Corollary 4.2.5.

We prove Items 4 and 5. Let us show that $A(t)$ converges in U_Γ to a flat connection Γ'. By compactness of the character variety and weak convergence, it suffices to show that if t_n' and t_n'' are sequences going to infinity and if $A(t_n') \mapsto \Gamma'$ and $A(t_n'') \mapsto \Gamma''$ for flat connections Γ' and Γ'', then $\Gamma' = \Gamma''$. Without loss of generality we can assume that $t_n' \leq t_n''$ for all n. Since ω is of finite energy with respect to g, we have

$$lim_{n\mapsto\infty}\int_{t_n'}^{t_n''}\int_N (|F_\omega|_g)^2 = 0.$$

Thus, according to Theorem 4.2.4 and Inequality 41, it follows that

$$lim_{n\mapsto\infty}||A(t_n') - A(t_n'')||_{L^2} = 0.$$

Also by 3.5.1

$$lim_{n\mapsto\infty}||F(A(t_n'))||_{L^2_{\frac{1}{2}}} = lim_{n\mapsto\infty}||F(A(t_n''))||_{L^2_{\frac{1}{2}}} = 0.$$

Thus, according to Lemma 4.1.1 there are flat connections Γ'_n and Γ''_n such that

$$lim_{n\mapsto\infty}||A(t'_n) - \Gamma'_n||_{L^2_{\frac{3}{2}}} = 0$$

and

$$lim_{n\mapsto\infty}||A(t''_n) - \Gamma''_n||_{L^2_{\frac{3}{2}}} = 0.$$

Clearly, then $\Gamma' = lim_{n\mapsto\infty}\Gamma'_n$ and $lim_{n\mapsto\infty}\Gamma''_n = \Gamma''$. It follows immediately that $||\Gamma' - \Gamma''||_{L^2} = 0$, and hence that $\Gamma' = \Gamma''$. This completes the proof that there is a flat connection $\Gamma' \in U_\Gamma$ such that

$$lim_{t\mapsto\infty}||A(t) - \Gamma'||_{L^2_{\frac{3}{2}}} = 0.$$

Lastly, according to Theorem 2.6.3 we have

$$||\alpha(t)||_{L^2_{\frac{7}{4}}} \leq \frac{1}{8}||\nabla_{A(t)} f_\Gamma||_{L^2}.$$

Since $lim_{t\to\infty}A(t) = \Gamma'$ in U_Γ where Γ' is a flat connection, it is clear from Item 2 of Lemma 2.5.1 that $lim_{t\to\infty}||\nabla_{A(t)} f_\Gamma||_{L^2} = 0$. □

DEFINITION 4.3.2. *Suppose that V_Γ is a coordinate patch centered at a C^∞-flat connection Γ. We say that V_Γ is an (ϵ_0, θ)-*Simon coordinate patch *if there is a gradient flow coordinate patch U_Γ centered at a Γ with $V_\Gamma \subset U_\Gamma$ so that the pair (V_Γ, U_Γ) satisfies Theorem* 4.3.1 *with respect to the constants ϵ_0 and* $0 < \theta \leq 1/2$. *An ϵ_0-Simon patch is an (ϵ_0, θ)-Simon patch for some* $0 < \theta \leq 1/2$. *Furthermore, the pair $V_\Gamma \subset U_\Gamma$ is called an ϵ_0-Simon patch pair.*

From this local result and the compactness of the character variety, we find that a neighborhood of $\chi(N) \subset \mathcal{B}_{\frac{3}{2}}(N)$ is covered by the images of (ϵ_0, θ)-Simon coordinate patches in $\mathcal{A}_{\frac{3}{2}}(N)$. More precisely, we have the following corollary.

COROLLARY 4.3.3. *There are constants $\epsilon_0 > 0$ and $0 < \theta \leq 1/2$ such that for any $g \in B$, for all $1 \leq T_0 \leq T_1 \leq \infty$ and any $L^2_{2,loc}$ g- ASD connection ω of finite* *energy, if*

$$\int_{T_0-1}^{T_1+1}\int_N (|F_\omega|_g)^2 < \epsilon_0$$

then there is a gradient flow patch U_Γ *and an* $L^2_{3,loc}$*-bundle isomorphism* $\varphi\colon [T_0,T_1)\times\eta \to E|([T_0,T_1)\times N)$ *such that* $\varphi^*\omega = A(t)+\alpha(t)dt$ *is a* C^3 *connection in standard form with respect to* U_Γ *and so that Inequalities* 46 *and* 47 *hold.*

The constant ϵ_0 depends only on the oriented riemannian manifold N. For the rest of this paper we fix $\epsilon_0 > 0$ so that Corollary 4.3.3 holds, and we define a Simon patch or a Simon patch pair to be an ϵ_0-Simon patch or an ϵ_0-Simon patch pair.

DEFINITION 4.3.4. *Let* $\mathcal{S}_\Gamma$ *be a pair* (V_Γ, T_0) *where* V_Γ *is a Simon patch and* $T_0 \geq 1$. *We call such triples Simon data centered at* Γ. *(If* Γ *is clear from the context we may drop it from the notation.) For any metric* $g\in B$*, we define*

$$\mathcal{M}_{\ell,w}(\mathcal{S}_\Gamma, g)\subset \mathcal{M}_{\ell,w}(X,g)$$

to be the subset of all classes of g*-ASD connections* ω *of energy* $2\pi^2\ell$ *satisfying*

(i) $\int_{T_0-1}^{\infty}\int_N (|F_\omega|_g)^2 < \epsilon_0$, *and*
(ii) $\omega\Big|(E|\{T_0\}\times N)$ *is* $L^2_{\frac{5}{2}}$*-isomorphic to a connection* $A_0\in V_\Gamma$.

Similarly, we have the based version $\mathcal{M}^0_{\ell,w}(\mathcal{S}_\Gamma, g)\subset\mathcal{M}^0_{\ell,w}(X,g)$.

For each set of Simon data $\mathcal{S}_\Gamma$ the subset $\mathcal{M}_{\ell,w}(\mathcal{S}_\Gamma, g)$ is an open subset of $\mathcal{M}_{\ell,w}(X,g)$.

Theorem 4.3.1 tells us that for any $[\omega]\in\mathcal{M}_{\ell,w}(\mathcal{S},g)$ the restriction of the connection ω to $E|([T_0,\infty)\times N)$ is $L^2_{3,loc}$-isomorphic to a connection of the form $A(t)+\alpha(t)dt$ where $A(t)$ is a path of finite length in a gradient flow patch U_Γ centered at Γ with a limit point Γ' which is a flat connection.

4.4. Proof of Theorem 4.0.1

Now we turn to the proof of Theorem 4.0.1. Throughout this section and the next we fix a C^3-metric g on X contained in B.

Let ω be a finite energy $L^2_{2,loc}$ g-ASD connection on E. By Theorem 4.3.1 there is a flat connection Γ' and an $L^2_{3,loc}$-bundle isomorphism φ so that the conclusions of Theorem 4.3.1 hold for $\varphi^*\omega = A(t) + \alpha(t)dt$ and Γ'. It is clear from the fact that $\mathcal{B}_{\frac{3}{2}}(N)$ is a Hausdorff space that assigning to each g- ASD connection ω the equivalence class of such a Γ' defines function

$$\partial\colon \mathcal{M}_w(X,g) \to \chi(N).$$

The only thing remaining to prove in order to establish Theorem 4.0.1 is the continuity of this function. By Corollary 4.1.3 it suffices to consider connections with a fixed energy. Suppose that $[\omega] \in \mathcal{M}_{\ell,w}(X,g)$. Let Γ be a C^∞ flat connection on η with

$$\partial([\omega]) = [\Gamma].$$

Fix an open neighborhood $\mathcal{N}$ of $\partial([\omega])$ in $\mathcal{B}_{\frac{3}{2}}(N)$. We shall find an open subset $W \subset \mathcal{M}_{\ell,w}(X,g)$ containing $[\omega]$ such that $\partial(W) \subset \mathcal{N}$. First, choose a gradient flow patch U_Γ centered at Γ such that the image of U_Γ in $\mathcal{B}(N)$ is contained in $\mathcal{N}$. According to Theorem 4.3.1 there is a Simon coordinate patch pair $V_\Gamma \subset U_\Gamma$. By the fact that ω has finite energy the fact that $\partial([\omega]) = \Gamma$, there is $T_0 \geq 1$ such that $[\omega] \in \mathcal{M}_{\ell,w}(\mathcal{S},g)$ where $\mathcal{S}$ is the pair (V_Γ, T_0). We claim that $\partial(\mathcal{M}_{\ell,w}(\mathcal{S},g)) \subset \mathcal{N} \cap \chi(N)$. By Theorem 4.2.4 for any $[\omega'] \in \mathcal{M}_{\ell,w}(\mathcal{S},g)$ there is an $L^2_{3,loc}$-bundle isomorphism $\varphi'\colon [0,\infty)\times\eta \to E|[0,\infty)\times N$ such that $(\varphi')^*\omega' = A'(t)+\alpha'(t)dt$ with $A'(t)$ in U_Γ for all $t \geq T_0$ and with $lim_{t\to\infty}A'(t) = \Gamma''$ where Γ'' is a flat connection in U_Γ. This proves that $\partial([\omega']) = \Gamma'' \in \mathcal{N} \cap \chi(N)$. Since the natural topology on $\chi(N)$ agrees with the subspace topology from $\mathcal{B}_{\frac{3}{2}}(N)$ this proves Theorem 4.0.1.

4.5. Existence of limit connection implies finite energy

Though the material in this section is not logically necessary for our arguments, we have decided to include it in order to round out the picture of L^2 ASD connections. The result is that on a cylindrical end manifold any ASD connection with a well-defined limit at infinity is in fact an L^2-connection.

PROPOSITION 4.5.1. *Let g be a cylindrical end metric on X and let ω be an g-ASD $L^2_{2,loc}$ connection on E with the property that*

$$lim_{t\mapsto\infty}[\omega|\{t\}\times N]$$

exists in $\mathcal{B}(N)$. Then ω is a finite energy connection.

Proof. The path $A_t = [\omega|\{t\}\times N] \in \mathcal{B}(N)$ lifts to a path $\tilde{A}(t)$ in $\tilde{\mathcal{B}}(N)$. Clearly, since $\tilde{\mathcal{B}}(N) \to \mathcal{B}(N)$ is a covering space, the path $\tilde{A}(t)$ has a limit as $t \mapsto \infty$. Of course, by Equation 2 and the fact that ω is ASD we have

$$\int_{T_0}^{T_1}\int_N |F_\omega|^2 = CS(\tilde{A}(T_1)) - CS(\tilde{A}(T_0)).$$

The finiteness of the energy of ω follows immediately. $\square$

4.6. The moduli space based at infinity

So far we have been working in the unbased context. Near a reducible it is convenient to have at our disposal the based moduli space and to work equivariantly. Let us assume that N is connected. Fix a base point $(0,n) \in \{0\}\times N \subset X$. The moduli space based at infinity is defined in terms of pairs $\{(\omega, e)\}$ where ω is a $L^2_{2,loc}$ g-ASD connection of finite energy and $e \in E|(0,n)$. We form the quotient space $\mathcal{M}^0_{\ell,w}(X,g)$ of these pairs by the action of the group of $L^2_{3,loc}$ gauge transformations. For each set of Simon data $\mathcal{S}$ we have the open subset $\mathcal{M}^0_{\ell,w}(\mathcal{S}) \subset \mathcal{M}^0_{\ell,w}(X,g)$. It is not completely obvious that the boundary map has a lift to an G-equivariant map. Since ω restricts to an $L^2_{\frac{1}{2}}$-loc connection on $[0,\infty)\times\{n\}$, given the pair (ω, e) there is a unique $L^2_{\frac{3}{2}}$-path $e(t)$, $0 \le t < \infty$, with $e(t) \in E|(t,n)$ and $e(t)$ parallel under ω. At issue is whether or not this path has a limit as $t \to \infty$; the equivariance will follow from the uniqueness of the limit. In order to ensure that this path has a limit we will assume that the metric g has exponential decay to a fixed metric g_0.

THEOREM 4.6.1. *Fix $\mu \geq 1$. Suppose that g is a metric in B which for $0 \leq t < \infty$ satisfies*

$$||g(t,n) - g_0(t,n)||_{C^1} < e^{-\mu t}. \tag{55}$$

Let $\varphi\colon [0,\infty)\times\eta \to E|([0,\infty)\times N)$ be a $L^2_{3,loc}$ bundle isomorphism. For any equivalence class $[\omega, f] \in \mathcal{M}^0_{\ell,w}(X,g)$ the associated path $[A(t), e(t)] = [\varphi^*\omega|\{t\}\times\eta, \varphi^{-1}(f(t))] \in \mathcal{B}^0_{\frac{3}{2}}(N)$ has a limit $[\Gamma', e_\infty]$. Associating to $[\omega, e]$ the element $[\Gamma', e_\infty]$ defines a continuous G-equivariant map

$$\partial^0 : \mathcal{M}^0_{\ell,w}(X,g) \to \mathcal{R}(N)$$

where $\mathcal{R}(N)$ is the representation variety of N. The quotient of ∂^0 modulo the G-action is the map $\partial\colon \mathcal{M}_{\ell,w}(X,g) \to \chi(N)$.

The proof of this theorem is based on the following lemma.

PROPOSITION 4.6.2. *Let V_Γ be a Simon patch and let $g \in B$ satisfying Inequality 55. There exists a constant C' depending only on V_Γ such that for all $T_0 \geq 1$ if $\omega = A + \alpha dt$ is a C^3 finite energy g-ASD connection on $[T_0 - 1, \infty)\times\eta$ in standard form with respect to V_Γ satisfying Inequality 49 with $T_1 = \infty$, then*

$$\int_{T_0}^{\infty} |\alpha(t)(n)| < C'\left(\int_{T_0-1}^{\infty}\int_N (|F_\omega|_g)^2\right)^{\theta}. \tag{56}$$

Proof. Let $V_\Gamma \subset U_\Gamma$ be a Simon patch pair centered at Γ. Then $A(t) \in U_\Gamma$ for $T_0 - 1 \leq t < \infty$. By Inequalities 27 and 24 and the fact that U_Γ is a gradient flow patch we have

$$\begin{aligned} ||\alpha(t)||_{C^0} &\leq ||\alpha(t) - \Theta(A(t))||_{C^0} + ||\Theta(A(t))||_{C^0} \\ &\leq K_1||h(t)||_{C^1}||F_{A(t)}||_{L^2_1} + K_0||a||_{L^2_{\frac{3}{2}}}||\nabla_{A(t)} f_\Gamma||_{L^2} \end{aligned}$$

By Lemma 3.5.1 this implies that

$$||\alpha(t)||_{C^0} \leq C||\nabla_{A(t)} f_\Gamma||_{L^2} + Ce^{-\mu t}||F_\omega||_{L^2([T_0-1,\infty)\times N)}.$$

for some constant $C > 0$ depending only on U_Γ and B. But for $g \in B$ we have $\|\nabla_{A(t)} f_\Gamma\|_{L^2} \leq 2\|\frac{\partial A(t)}{\partial t}\|_{L^2}$ for all $t \geq T_0 - 1$. Integrating this inequality from T_0 to infinity and invoking Part 3 of Theorem 4.3.1 gives

$$\int_{T_0}^{\infty} |\alpha(t)(n)| \leq K \left(\int_{T_0-1}^{\infty} \int_N (|F_\omega|_g)^2 \right)^\theta + \mu C' \|F_\omega\|_{L^2(T_0-1,\infty)\times N}.$$

Since $\theta \leq 1/2$, the result follows immediately. □

Now let us begin the proof of Theorem 4.6.1.

Proof. Let us first prove that the limit exists. Let ω be an $L^2_{2,loc}$ finite energy g-ASD connection, and let Γ be a C^∞-flat connection such that $\partial([\omega)] = [\Gamma]$. By Theorem 4.3.1 we can find a set of Simon data $\mathcal{S}$ so that $[\omega] \in \mathcal{M}_{\ell,w}(\mathcal{S})$. Let $\varphi\colon [T_0,\infty) \times \eta \to E\Big|([T_0,\infty) \times N)$ be an $L^2_{3,loc}$ bundle isomorphism for which the conclusions of Theorem 4.3.1 hold.

Let $\tilde{e}$ be an ω-parallel frame over $[0,\infty) \times \{n\}$. Using φ we write $\tilde{e} = e(t)$ where $e(t) \in \eta|\{n\}$. Let $\sigma(t) \in G$ be the path such that $e(t) = e(0) \cdot \sigma(t)$. For $e(t)$ to be parallel under ω means that

$$\begin{aligned} \sigma(t)^{-1}\dot{\sigma}(t) &= -(\alpha(t))(n) \quad with \\ \sigma(0) &= Id \end{aligned}$$

Since $\alpha(t,n)$ is a C^3 function of t, there is a unique C^4-solution to this equation. Since

$$|\dot{\sigma}(t)| = \big|(\alpha(t))(n)\big|$$

it follows from Lemma 4.6.2 that the path $\sigma(t)$ has finite length and hence $\sigma_\infty = lim_{t\to\infty}\sigma(t)$ exists. Set $e_\infty = e(0) \cdot \sigma_\infty$. Clearly, we have

$$lim_{t\to\infty}[A(t), e(t)] = [\Gamma, e_\infty].$$

Associating $[\Gamma, e_\infty]$ to $[\omega]$ defines an G-invariant function

$$\partial^0 : \mathcal{M}^0_{\ell,w}(X,g) \to \mathcal{R}(N)$$

covering the map ∂ in Theorem 4.0.1. Arguing as in the proof of Theorem 4.0.1 the continuity of ∂^0 is easily established using the uniform estimates given in Lemma 4.6.2 on the length of $\alpha(t)(n)$. □

DEFINITION 4.6.3. *For any subset $K \subset \mathcal{R}(N)$ we denote by $\mathcal{M}^0_{\ell,w}(X,g,K)$ the subset $(\partial^0)^{-1}(K) \subset \mathcal{M}^0_{\ell,w}(X,g)$.*

4.6.1. The case when N is not connected. The above discussion requires some minor modifications when N is not connected. Fix a collection of base points $\{n^1, ..., n^k\}$, one in each component of N. By the representation variety $\mathcal{R}(N)$ we mean the product of the representation varieties of the components. This of course is identified with the space of gauge equivalence classes of based flat connections on N, based at the points $\{n^1, \dots, n^k\}$. The quotient of $\mathcal{R}(N)$ by the action of conjugation by G^k is the character variety $\chi(N)$ and is identified with the space of gauge equivalence classes of flat connections on N.

We defined the moduli space based at infinity to be the set of pairs $\{(\omega, (e_1, \dots, e_k))\}$ where ω is a $L^2_{2,loc}$ g-ASD connection of finite energy and $e_j \in E|(0, n^j)$. We form the quotient space $\mathcal{M}^0_{\ell,w}(X,g)$ of these pairs by the action of the group of $L^2_{3,loc}$ gauge transformations.

With these definitions the statement and proof of Theorem 4.6.1 remain valid without the assumption that N is connected.

4.7. Uniform bounds

We end this section with a useful technical result which describes properties of connections in a neighborhood of point in $\mathcal{M}_{\ell,w}(\mathcal{S},g)$.

LEMMA 4.7.1. *Let $g \in B$ and let Γ be a C^∞-flat connection on η. Let $\mathcal{S} = (V_\Gamma, T_0)$ be a set of Simon data and $V_\Gamma \subset U_\Gamma$ a Simon patch pair. For any $0 < \epsilon \leq \epsilon_0$ any $\omega_1 \in \mathcal{M}_{\ell,w}(\mathcal{S},g)$ there is a $T_1 > T_0$ so that for any $\delta > 0$ there is a neighborhood $\mathcal{O}$ of ω_1 in $\mathcal{M}_{\ell,w}(\mathcal{S},g)$ so that for any $\omega_2 \in \mathcal{O}$ the following holds. For $i = 1, 2$*

there are bundle isomorphisms φ_i which put ω_i in standard form with respect to Γ and C^∞-flat connections $\Gamma_i \in U$ such that the following holds. For any $T > T_0$

$$||\varphi_1^* \omega_1 - \varphi_2^* \omega_2||_{L_3^2([T-1,T+1]\times N)} < \delta.$$

Furthermore there are C^∞ flat connections Γ_1 and Γ_2 so that for any $T > T_1$ and $i = 1$ or 2 we have

$$||\varphi_i^* \omega_i - \Gamma_i||_{L_3^2([T-1,T+1]\times N)} < \epsilon$$

and

$$\int_{[T_1,\infty)\times N} (||F_{\omega_2}||_g)^2 \leq \epsilon.$$

Proof. For any constant $\delta_1 > 0$ we can find a compact set $K \subset X$ so that $\int_{X-K} tr(F_{\omega_1} \wedge F_{\omega_1}) < \delta_1$. Let $\mathcal{O}$ be the open subset of $\mathcal{M}_{\ell,w}(\mathcal{S}, g)$ consisting of connections ω so that

$$||\omega - \omega_1||_{L_3^2(K)} < \delta_2.$$

Then there is a constant $C > 0$ so that for any ω satisfying the above inequality we have $\ell - \int_K tr(F_\omega \wedge F_\omega) < \delta_1 + C\delta_2$. There is a $T_1 > T_0$ with $([T_1, \infty) \times N) \cap K = \emptyset$. For any $T > T_1$ Theorem 4.3.1 applies for the same Simon patch V_Γ . The uniqueness assertion in Theorem 4.3.1 guarantees that bundle isomorphisms will be the same (up to pullback on an element in Stab(Γ)) wherever they are defined. Thus $\lim_{t\to\infty} A_2(t)$ exists and we let its limit be Γ'. Furthermore we have for all $T > T_1$

$$||A_2(T) - \Gamma'||_{L^2(N)} < (\delta_1 + C\delta_2)^\theta.$$

Hence $||A(t) - \Gamma'||_{L^2([T-1,T+1]\times N)} < 2(\delta_1 + C\delta_2)^\theta$ Lemma 3.3.2 now applies for δ_1, δ_2 small enough to give a constant $C_1 > 0$ so that

$$||A(t) - \Gamma'||_{L_3^2([T-1,T+1]\times N)} < 2C_1(\delta_1 + C\delta_2)^\theta.$$

If we take δ_1, δ_2 small enough we can arrange that the desired estimate holds. □

CHAPTER 5

The finite dimensional ODE dominating the Chern-Simons flow

In Theorem 2.6.3 we showed that, for any metric g on X sufficiently close to g_0, the g-ASD equation for a connection in a bundle $[0,\infty)\times\eta$ is equivalent to a perturbation of the gradient flow equation for the restriction of the Chern-Simons function to a slice S_Γ in $\mathcal{A}(\eta)$. We then deduced the existence of an asymptotic flat connection for any finite energy g-ASD connection over X. In this chapter we produce a refinement of this result. We show that near Γ this perturbation of the gradient flow equation for f_Γ in S_Γ is dominated by the restriction of the gradient flow equation to an invariant finite dimensional sub-manifold $\mathcal{H}_\Gamma \subset S_\Gamma$. This submanifold is tangent to the null-space of the Hessian of f_Γ at $\Gamma \in S_\Gamma$. In Lemmas 2.1.1 and 2.5.1 we identified this Hessian with $-*d_\Gamma|Ker d^* \cap \Omega^1(N)$. Thus, $\mathcal{H}_\Gamma$ is tangent at Γ to $H^1(N; ad\eta)$. The submanifold $\mathcal{H}_\Gamma$ is called *the center manifold*. We show that analogously to what happens in the finite dimensional case, all finite energy flow lines in S_Γ approach exponentially fast to a flow line in $\mathcal{H}_\Gamma$. The exponent in this approach is the first positive eigenvalue μ_Γ^+ of the Hessian of f_Γ, that is to say of $-*d_\Gamma|Ker\, d_\Gamma$.

5.1. Existence of the center manifold

We deduce the existence of the center manifold in our context from a general theorem about center manifolds for flows on Banach spaces. The following is a rephrasing of Theorem A.1 of [11], where we have made the extra simplifying assumptions that the Banach space in question is a Hilbert space and that the linear map in the center direction (H^0) is trivial.

THEOREM 5.1.1. *Let H be a Hilbert space. Suppose that*

$$H = H^+ \oplus H^0 \oplus H^-$$

is a decomposition of H. Let

$$L^+ : H^+ \to H^+ \ ; \qquad L^- : H^- \to H^-$$

be densely defined, closed, unbounded Fredholm linear operators. Suppose that for each $t \geq 0$ the maps

$$e^{-L^+t} \colon H^+ \to H^+ \ ; \qquad e^{L^-t} \colon H^- \to H^-$$

are bounded linear maps, and that these define semi-groups of endomorphisms of H^+ and H^- which vary continuously with t in the strong operator topology. Suppose that there exist positive constants λ_+, λ_-, D such that

$$sup_{t\geq 0} max\left(e^{\lambda_+ t}||e^{-L^+t}||, e^{\lambda_- t}||e^{L^-t}||\right) \leq D \tag{57}$$

where the norms are the operator norms on bounded linear operators on $H^{\pm}$. Let $L \colon H \to H$ be the densely defined linear operator given by $L(x^+, x^0, x^-) = L^+(x^+) + L^-(x^-)$.

Let $U \subset H$ be a neighborhood of $\{0\}$ and let $n \colon U \to H$ be a smooth function which vanishes to second order at $\{0\} \in H$. Consider the densely defined vector field Ξ on U given by

$$\Xi(x) = L(x) + n(x).$$

Then for any $m \geq 1$ there is a connected neighborhood $\mathcal{N} \subset H^0$ of $\{0\}$ and a C^m map $\zeta \colon \mathcal{N} \to H^+ \oplus H^-$ whose graph $\mathcal{H}$ satisfies:

(1) $\mathcal{H} \subset U$,

(2) *$\mathcal{H}$ is tangent to H^0 at $\{0\}$,*

(3) *$\mathcal{H}$ is contained in the domain of definition of the vector field Ξ,*

(4) *Ξ is everywhere tangent to $\mathcal{H}$,*

(5) *every critical point of Ξ sufficiently close to $\{0\}$ is contained in $\mathcal{H}$.*

DEFINITION 5.1.2. *Any submanifold $\mathcal{H}$ as in the above theorem is called a C^m-center manifold for the flow Ξ, or a center manifold for Ξ for short.*

The center manifold is not locally unique. To construct it one must extend the densely defined flow Ξ on a neighborhood of $\{0\}$ to a flow $\Xi_{global} = L + n_{global}$ which is densely defined on the entire space, which agrees with the linear flow L outside of a small neighborhood of $\{0\}$, and for which the non-linear term n_{global} is everywhere small with respect to the linear term. Once such an extension Ξ_{global} is chosen, the center manifold for it exists globally and is unique. To construct such an extension in our case when the underlying space is a Hilbert space one chooses a weakly monotone decreasing C^∞-function $\varphi \colon [0,\infty) \to [0,1]$ with $\varphi = 1$ near $\{0\}$ and $\varphi(t) = 0$ for $t \geq 1$. The (densely defined) global vector is given by $\Xi_{global}(x) = L(x) + \varphi(\lambda||x||)n(x)$. For all λ sufficiently large this flow will have the requisite properties.

Just as in the finite dimensional case there are also 'invariant' submanifolds W^{cs} and W^{cu} tangent respectively to $H^+ \oplus H^0$ and $H^- \oplus H^0$. These submanifolds are called *center-stable* and *center-unstable* manifolds. The center-stable manifold W^{cs} is obtained as a graph of a C^m-map $\zeta^+ \colon H^+ \oplus H^0 \to H^-$.

Their intersection is a center manifold. On W^{cs} the vector field Ξ is defined on the image of the domain of L^+ on the subset where it is defined, it is tangent to W^{cs}. Similarly, for W^{cu}. The proof of the existence of the center manifold goes by first constructing W^{cs} and W^{cu} and then taking their intersection.

One consequence of this description is equivariance for the center manifold.

COROLLARY 5.1.3. *With $H, U, L,$ and n as in the statement of Theorem 5.1.1. Suppose that on some neighborhood of $\{0\}$ in H the functions L and n are invariant under a group S of isometries of H fixing $\{0\}$. Then there is a center manifold $\mathcal{H}$ for Ξ which is invariant under S.*

Now we will show that this theorem applies to our context. Let $\Omega^1_2(N) \cap Ker\, d^*_\Gamma = \mathcal{K}_\Gamma \cap L^2_2$ and let $L = -*\,d_\Gamma$. Let H^+, H_Γ, and H^- be the closures of the positive, zero, and negative eigenspaces of L. Then according to Part 2 of Lemma 2.1.1 $H = H^+ \oplus H_\Gamma \oplus H^-$ is an L^2-orthogonal decomposition of H. We denote by $H^\perp_\Gamma$ the direct sum $H^+ \oplus H^-$.

We claim that e^{-L^+t} and e^{L^-t} define semi-groups of bounded endomorphisms of H^+ and H^-, respectively, which vary continuously with t in the strong operator topology. Let us consider e^{-L^+t} on H^+. According to Part 2 of Lemma 2.1.1 any element in H^+ can be written as an infinite linear combination

$$\sum_{\lambda^+} \alpha_{\lambda^+}$$

where λ^+ ranges over the positive eigenvalues of L and for each λ^+ the element α_{λ^+} is a eigenvector for L with eigenvalue λ^+. In general, such an infinite sum defines an element of H if and only if

$$\sum_{\lambda^+} ||\alpha_{\lambda^+}||^2 < \infty.$$

By definition,

$$e^{-L^+t}(\alpha_{\lambda^+}) = e^{-\lambda^+t}\alpha_{\lambda^+}.$$

This obviously defines a semi-group of endomorphisms on the span of the eigenspaces of L^+. Thus, to show that e^{-L^+t} is semi-group of bounded endomorphisms of H^+

we need only see that for each $t \geq 0$ we have

$$\sum_{\lambda+} \left(e^{-\lambda^+ t}||\alpha_{\lambda+}||\right)^2 < B(t) \sum_{\lambda+} ||\alpha_{\lambda+}||^2$$

for some constant $B(t)$ depending on t. Since the λ^+ are all bounded below by μ_Γ^+, this is obvious. Similar arguments show that these operators vary continuously with t in the strong operator topology.

These arguments apply *mutatis mutandis* to show that $e^{L^- t}$ defines a semi-group of bounded endomorphisms on H^- continuous in the strong operator topology. These computations also show that Condition 57 holds for $\lambda_+ = \mu_\Gamma^+$, $\lambda_- = \mu_\Gamma^-$, and $D = 1$. Let $U_\Gamma \subset S_\Gamma$ be a coordinate patch centered at Γ for which Lemma 2.5.1 holds. Let $U \subset \mathcal{K}_\Gamma$ be such that $U_\Gamma = \{\Gamma\} + U$. As in Lemma 2.5.1, we have a function $n\colon U \cap L_2^2 \to H$ given by

$$n(a) = *a \wedge a + d_A(\theta(A))$$

where $A = \Gamma + a$. By Part 4 of Lemma 2.5.1 the function n is a smooth function from $\mathcal{U} \cap L_2^2$ to H vanishing to second order at $\{0\}$. According to Lemma 2.5.1,

$$\nabla_{\Gamma+a} f_\Gamma = - * d_\Gamma(a) + n(a).$$

Clearly, the function $f_\Gamma = CS|S_\Gamma$ is invariant under $Stab(\Gamma)$. Also, the metric on $\mathcal{K}_\Gamma$ with which respect to which we are taking the gradient is invariant under $Stab(\Gamma)$. Thus, $\nabla_{\Gamma+a} f_\Gamma$ is a $Stab(\Gamma)$- invariant vector field on $\mathcal{K}_\Gamma$.

COROLLARY 5.1.4. *Let Γ be a C^∞-flat connection on η. Let $U_\Gamma \subset S_\Gamma$ be a coordinate patch centered at Γ for which Lemma 2.5.1 holds. Let $U \subset \mathcal{K}_\Gamma$ be given by $\{\Gamma\} + U = U_\Gamma$. Let $H_\Gamma^\perp \subset \mathcal{K}_\Gamma$ be the L^2-orthogonal complement of $Ker(-*d_\Gamma|\mathcal{K}_\Gamma) = H^1(N; ad\,\Gamma)$. There is a $Stab(\Gamma)$-invariant neighborhood $\mathcal{N}_\Gamma \subset H^1(N; ad\Gamma)$ of $\{0\}$ and a $Stab(\Gamma)$-invariant C^2-function $\zeta\colon \mathcal{N}_\Gamma \to H_\Gamma^\perp \cap L_2^2$ whose graph $\mathcal{H}_\Gamma$ is a $Stab(\Gamma)$- invariant C^2-center for the densely defined vector field $a \mapsto \nabla_{\Gamma+a} f_\Gamma$ on $U \cap L_2^2$.*

Proof. This is immediate from Theorem 5.1.1 and the above discussion. □

We can now introduce the set which provides the finite dimensional model for the L^2-moduli space. Fix a C^∞ flat connection Γ and a $Stab(\Gamma)$ equivariant center manifold $\mathcal{H}_\Gamma$. Let Ξ denote the restriction of the vector field ∇f_Γ to $\mathcal{H}_\Gamma$. For $h \in \mathcal{H}_\Gamma$ let h_t denote the maximal flow line of Ξ with initial condition $h_0 = h$. The flow need not be defined for all time.

DEFINITION 5.1.5. *Let Y_Γ be the set of $h \in \mathcal{H}_\Gamma$ so that h_t is defined for all $t \geq 0$ and $\lim_{t\to\infty} h_t$ exists and is in $\mathcal{H}_\Gamma$. The set Y_Γ is called the stable set of Ξ.*

5.2. Exponential approach to the center manifold: the statement

Next we show that corresponding to any finite energy flow line in S_Γ coming from a g-ASD connection there is an exponentially asymptotic flow line in the center manifold.

Recall from the last section that we have a $Stab(\Gamma)$ invariant decomposition

$$\mathcal{K}_\Gamma = H_\Gamma \times H_\Gamma^\perp$$

where $H_\Gamma = H^1(N; ad\Gamma)$ and $H_\Gamma^\perp$ is the L^2-orthogonal complement of H_Γ. We denote by Π^0 and $\Pi^\perp$ the L^2-orthogonal projections of $\mathcal{K}_\Gamma$ onto H_Γ and $H_\Gamma^\perp$, respectively. We also have a $Stab(\Gamma)$ invariant C^2-center manifold $\mathcal{H}_\Gamma \subset \mathcal{K}_\Gamma \cap L^2_2$ described as the graph of a $Stab(\Gamma)$ equivariant C^2-function ζ from an open neighborhood $\mathcal{N}_\Gamma$ of $\{0\}$ in H_Γ to $H_\Gamma^\perp \cap L^2_2$.

It is convenient to state our result in terms of connections in standard form. To this end we make the following definition.

DEFINITION 5.2.1. *Given Simon data $\mathcal{S}_\Gamma = (V_\Gamma, T_0)$ and any metric $g \in B$, for any $\epsilon, 0 < \epsilon \leq \epsilon_0$ we define $Std(\mathcal{S}, g, \epsilon)$ to be the set of all C^3, g-ASD connections $\omega = A(t) + \alpha(t)dt$ on $[T_0 - 1, \infty) \times \eta$ satisfying*

(i) $\int_{T_0-1}^{\infty} \int_N (|F_\omega|_g)^2 < \epsilon$,

(ii) $A(T_0) \in V_\Gamma$.

(iii) ω *is in standard form with respect to* Γ.

$Std(\mathcal{S}, g, \epsilon)$ is given the topology of uniform convergence in the L^2_2 norm on compact sets together with the convergence of $\int_{T_0-1}^{\infty}\int_N (|F_\omega|_g)^2$. It follows from Theorem 4.3.1 that there is a coordinate patch U_Γ so that for all $t \geq T_0 - 1$ we have $A(t) \in U_\Gamma$ and that $\lim_{t\to\infty} A(t)$ exists and is a flat connection in U_Γ. Also note that $Stab(\Gamma)$ acts on $Std(\mathcal{S}, g, \epsilon)$. We denote $Std(\mathcal{S}, g, \epsilon_0)$ by $Std(\mathcal{S}, g)$.

Here is the main result on the existence of flow lines in the center manifold exponentially close to flow lines for finite energy ASD connections.

THEOREM 5.2.2. *Let* Γ *be a* C^∞*-flat connection on* η. *Let* $\mathcal{N}_\Gamma$ *be a* $Stab(\Gamma)$*-equivariant neighborhood of* $\{0\}$ *in* $H^1(N; ad\Gamma)$, *and let* $\zeta\colon \mathcal{N}_\Gamma \to H_\Gamma^\perp \cap L^2_2$ *be a* $Stab(\Gamma)$ *invariant* C^2*-function whose graph* $\mathcal{H}_\Gamma \subset \mathcal{N}_\Gamma \times H_\Gamma^\perp = Z_\Gamma \subset K_\Gamma$ *is a* $Stab(\Gamma)$ *invariant* C^2 *center manifold. Let* $\mu > \mu_\Gamma^-$. *Then there is Simon data* $\mathcal{S} = (V_\Gamma, T_0)$ *and* $U_\Gamma \subset Z_\Gamma$ *such that* $V_\Gamma \subset U_\Gamma$ *is a Simon patch pair and a constant* $\kappa > 0$ *so that the following holds. For any metric* $g \in B$ *with* $g = g_0 + h$ *satisfying* $||h(t)||_{C^1} \leq e^{-\mu t}$, *and* $\omega = A + \alpha dt \in Std(\mathcal{S}, g)$ *the following holds. There is a unique path* $B\colon [T_0 - 1, \infty) \to \mathcal{H}_\Gamma$ *so that*

$$\frac{\partial B}{\partial t} = \nabla_B f_\Gamma$$

so that forming the g_0*-ASD connection* $\tilde{\omega} = B(t) + \Theta(B(t))dt$, *we have for* $T_0 \leq t < \infty$

$$||\omega - \tilde{\omega}||_{L^2_3([t-\frac{1}{2}, t+\frac{1}{2}]\times N)} < \kappa e^{-\mu_\Gamma^-(t-T_0)/2}.$$

The assignment of $B(T_0)$ *to* ω *defines a continuous* $Stab(\Gamma)$*-equivariant function*

$$\tilde{\Upsilon}_{\mathcal{S}}\colon Std(\mathcal{S}, g) \to Y_\Gamma \subset \mathcal{H}_\Gamma.$$

Lastly the map sending $\omega \in Std(\mathcal{S}, g)$ *to* $\omega - \tilde{\omega}$ *defines a continuous* $Stab(\Gamma)$*-equivariant map* $Std(\mathcal{S}, g) \to \Omega^1_{2,\delta}([T_0, \infty) \times N)$.

This theorem allows us to define the local refinement of the boundary map $\partial^0 \colon \mathcal{M}^0_{\ell,w}(X,g) \to \mathcal{R}(N)$ described in the introduction.

Fix a set of Simon data $\mathcal{S}$ and let $[\omega, f] \in \mathcal{M}^0_{\ell,w}(\mathcal{S},g)$ be given. According to Theorem 4.3.1 there is a bundle isomorphism $\varphi \colon [T_0 - 1, \infty) \times \eta \to E|_{[T_0-1,\infty)\times N}$ so that $\varphi^*(\omega) = A(t) + \alpha(t)dt$ is in standard form with respect to Γ, in particular $\varphi^*(\omega) \in Std(\mathcal{S}, g)$. This bundle isomorphism is unique up to the pullback of an element of $Stab(\Gamma)$. Thus there is a well defined map

$$\Upsilon_{\mathcal{S}} \colon \mathcal{M}^0_{\ell,w}(\mathcal{S},g) \to Y_\Gamma \times_{Stab(\Gamma)} \eta_n$$

given by the equation

$$\Upsilon_{\mathcal{S}}([\omega, f]) = [\tilde{\Upsilon}_{\mathcal{S}}(\phi^*(\omega)), \phi^*(f)].$$

There is by definition a flow on the stable set Y_Γ and hence an induced flow $F_t \colon Y_\Gamma \times_{Stab(\Gamma)} \eta_n \to Y_\Gamma \times_{Stab(\Gamma)} \eta_n$. Furthermore for any equivalence class $[B, f] \in Y_\Gamma \times_{Stab(\Gamma)} \eta_n$ we have that $\lim_{t\to\infty} F_t([B,f])$ exists and is a gauge equivalence class of based flat connection and hence a point in $\mathcal{R}(N)$.

COROLLARY 5.2.3. *With same hypothesis and notation as in Theorem 5.2.2 the following holds. The map $\Upsilon_{\mathcal{S}} \colon \mathcal{M}^0_{\ell,w}(\mathcal{S},g) \to Y_\Gamma \times_{Stab(\Gamma)} \eta_n$ is continuous. Furthermore $\Upsilon_{\mathcal{S}}$ refines ∂^0 in that the following equation holds for all $[\omega, f] \in \mathcal{M}^0_{\ell,w}(\mathcal{S},g)$.*

$$\lim_{t\to\infty} F_t(\Upsilon_{\mathcal{S}}([\omega, f])) = \partial^0([\omega, f]).$$

The proof of Theorem 5.2.2 occupies the rest of this section.

The size of the neighborhood of $\{0\}$ that we can work with is governed by the following lemma and Inequality 19.

LEMMA 5.2.4. *Given $\mu > 0$ there is a neighborhood $\mathcal{N}'_\Gamma$ of $\{0\}$ in H_Γ so that for all $a \in \mathcal{N}'_\Gamma$ and all $\xi \in H_\Gamma$*

$$\|D_a\zeta(\xi)\|_{L^2} \le \|\xi\|_{L^2}. \tag{58}$$

Furthermore if we let $\mathcal{H}'_\Gamma$ denote the part of graph of ζ above $\mathcal{N}'_\Gamma$ then for all x, y in $\mathcal{H}'_\Gamma$ we have:

$$\|x - y\|_{L^2} \leq 2\|\Pi^0(x) - \Pi^0(y)\|_{L^2} \text{ and} \tag{59}$$

$$\|\nabla_x f_\Gamma - \nabla_y f_\Gamma\|_{L^2} \leq \frac{\mu_\Gamma^-}{4}\|\Pi^0(x) - \Pi^0(y)\|_{L^2}. \tag{60}$$

Proof. The first inequality is clear for $\mathcal{N}'_\Gamma$ sufficiently small since $\zeta|\mathcal{N}_\Gamma$ is a C^2-map whose differential at $\{0\}$ is zero. The second inequality is immediate from the first.

On each tangent space to $\mathcal{H}_\Gamma$ we take the restriction from K_Γ of the L^2-inner product. This is a C^1-riemannian metric on $\mathcal{H}_\Gamma$. The restriction of ∇f_Γ to $\mathcal{H}_\Gamma$ is a C^1 vector field. The Hessian of f_Γ at 0 vanishes. Thus, there is an inequality of the type desired in 60 except that the right hand side is an arbitrarily small constant times the riemannian distance in $\mathcal{H}_\Gamma$ between x and y. But the second inequality says that this distance is commensurate with $\|\Pi^0(x) - \Pi^0(y)\|_{L^2}$.

□

Now we fix a Simon patch pair $V_\Gamma \subset U_\Gamma \subset Y_\Gamma$ such that $(U_\Gamma) \subset (\Pi^0)^{-1}(\mathcal{N}'_\Gamma)$. There is no loss of generality in replacing $\mathcal{H}_\Gamma$ by $\mathcal{H}'_\Gamma$ and thus assuming that Lemma 5.2.4 holds for $\mathcal{H}_\Gamma$. We may also assume that for all flat connections Γ' in U_Γ is contained in the L^2_2 ball of radius $min\{\mu_\Gamma^-, \mu_\Gamma^+\}/32C_1$ about Γ where C_1 is the constant of Inequality 19. This is the Simon patch for which we will prove Theorem 5.2.2.

The existence of the center manifold allows us to decompose a connection $\omega \in Std(\mathcal{S}, g)$.

Definition 5.2.5. *A decomposition $\omega = \Gamma + b(t) + c(t) + \alpha(t)dt \in Std(\mathcal{S}, g)$ is said to be the standard decomposition (with respect to $\mathcal{H}_\Gamma$) if $b(t) \in \mathcal{H}_\Gamma$ and $c(t) \in H_\Gamma^\perp$.*

5.3. Approximate solutions in the center manifold

As a first step in proving Theorem 5.2.2 we prove a result about paths in the center manifold, showing that an approximate solution to the gradient flow equation is approximated by a true solution.

LEMMA 5.3.1. *Fix $C > 0$ and let $b\colon [T_0, \infty) \to \mathcal{H}_\Gamma$ be a C^2-path such that $\lim_{t\to\infty} b(t)$ exists and*

$$\|\dot{b}(t) - \nabla_{b(t)} f_\Gamma\|_{L^2} \le Ce^{-\mu_\Gamma^-(t-T_0)/2}. \tag{61}$$

Then there is a unique C^2-path $z\colon [T_0, \infty) \to \mathcal{H}_\Gamma$ with:

(1)

$$\dot{z}(t) = \nabla_{z(t)} f_\Gamma$$

and

(2)

$$\|z(t) - b(t)\|_{L^2} \le (8/\mu_\Gamma^-)Ce^{-\mu_\Gamma(t-T_0)/2}.$$

Proof. Define the Banach space $\overline{B}$ to be the space of continuous paths $\gamma\colon [T_0, \infty) \to H_\Gamma$ such that $\sup_{t\in[T_0,\infty)} e^{\mu_\Gamma^-(t-T_0)/2}\|\gamma(t)\|_{L^2} < \infty$. For $\gamma \in \overline{B}$ define

$$\|\gamma\|_{\overline{B}} = \sup_{t\in[T_0,\infty)} e^{\mu_\Gamma^-(t-T_0)/2}\|\gamma(t)\|_{L^2}.$$

For a path $\overline{b}\colon [T_0, \infty) \to H_\Gamma$ satisfying $\lim_{t\to\infty} \overline{b}(t) = \overline{b}(\infty)$, we define $J_{\overline{b}}\colon \overline{B} \to \overline{B}$ by the following equation

$$J_{\overline{b}}(\gamma)[t] = -\overline{b}(t) + \overline{b}(\infty) - \int_t^\infty \Xi\left(\overline{b}(s) + \gamma(s)\right) ds \tag{62}$$

where Ξ is the pullback under the obvious map of H_Γ onto $\mathcal{H}_\Gamma$ of the vector field ∇f_Γ on $\mathcal{H}_\Gamma$. Then Ξ is a C^1-vector field on H_Γ.

CLAIM 5.3.2. *Suppose that $b(t)$ satisfies the hypothesis of Lemma 5.3.1. We set $\overline{b}(t)$ equal to the path $\Pi_0(b(t))$. The map $J_{\overline{b}}\colon \overline{B} \to \overline{B}$ is a $\frac{1}{2}$-contraction. Furthermore, the unique fixed point β for this mapping satisfies*

$$||\beta(t)||_{L^2} < 4C/\mu_{\overline{\Gamma}} e^{-\mu_{\overline{\Gamma}}(t-T_0)/2}.$$

Proof. Suppose that a satisfies the hypothesis of Lemma 5.3.1. Since $||(\Pi_0)_*\tau||_{L^2} \leq ||\tau||_{L^2}$ for any tangent vector τ to $\mathcal{H}_\Gamma$, integration of equation 61 yields that $||J_{\overline{b}}(0)||_{\overline{B}} \leq 2C/\mu_{\overline{\Gamma}}$. We shall show that $J_{\overline{b}}$ is a $\frac{1}{2}$-contraction. We have

$$\begin{aligned}
&||J_{\overline{b}}(\gamma_1) - J_{\overline{b}}(\gamma_2)||_{\overline{B}} \\
&= \sup_{t\in[T_0,\infty)} e^{\mu_{\overline{\Gamma}}(t-T_0)/2}\left|\int_t^\infty (\Xi((b(s)+\gamma_1(s)) - \Xi((b(s)+\gamma_2(s)))\, ds\right| \\
&\leq \frac{\mu_{\overline{\Gamma}}}{4} \sup_{t\in[T_0,\infty)} e^{\mu_{\overline{\Gamma}}(t-T_0)/2}\int_t^\infty ||\gamma_1(s)-\gamma_2(s)||_{L^2} ds \\
&\leq \frac{\mu_{\overline{\Gamma}}}{4}||\gamma_1-\gamma_2||_{\overline{B}} \sup_{t\in[T_0,\infty)} e^{\mu_{\overline{\Gamma}}(t-T_0)/2}\int_{T_0}^\infty e^{-\mu_{\overline{\Gamma}}(s-T_0)/2} ds \\
&\leq \frac{1}{2}||\gamma_1-\gamma_2||_B.
\end{aligned}$$

Since the norm of $J_{\overline{b}}(0)$ is less than $2C/\mu_{\overline{\Gamma}}$ the fixed point of $J_{\overline{b}}$ has norm at most $4C/\mu_{\overline{\Gamma}}$, and thus the last inequality holds. □

Now we return to the proof of Lemma 5.3.1. Given $b(t)$ satisfying the hypothesis of that lemma, we let $\beta(t)$ be the unique fixed point of $J_{\overline{b}}$ on the ball $B(0, 4C/\mu_{\overline{\Gamma}})$. Set $\overline{z}(t) = \overline{b}(t) + \beta(t)$. Clearly, from the fact that $J_{\overline{b}}(\beta) = \beta$ by differentiating Equation 62 we see that $\Xi(\overline{z}) = \frac{\partial \overline{z}}{\partial t}$. Set $z(t)$ equal to the corresponding curve in $\mathcal{H}_\Gamma$. Then clearly, we have

$$\nabla_{z(t)} f_\Gamma = \frac{\partial z}{\partial t}.$$

Since the vector field ∇f_Γ is C^1, this implies that $z(t)$ is C^2. By Claim 5.3.2 we have $||\overline{z}(t) - \overline{b}(t)||_{L^2} \leq 4C/\mu_{\overline{\Gamma}} e^{-\mu_{\overline{\Gamma}}(t-T_0)/2}$. By assumption $\mathcal{H}_\Gamma$ satisfies Inequality 59. It follows immediately that $z(t)$ satisfies the second condition stated in the lemma.

The uniqueness assertion follows directly from the uniqueness of the fixed point of $J_{\bar{b}}$ □

As a corollary of this lemma we have that distinct flow lines in the center manifold cannot approach each other exponentially fast. More precisely.

COROLLARY 5.3.3. *Let $z_1, z_2 \colon [T_0, \infty) \to \mathcal{H}_\Gamma$ be flow lines for ∇f_Γ. If there is a constant C so that $||z_1(t) - z_2(t)||_{L^2} < Ce^{\frac{-\mu_\Gamma^-}{2}(t-T_0)}$ for all $t \geq T_0$ then $z_1(t) = z_2(t)$.*

Proof. This follows directly from the uniqueness assertion of Lemma 5.3.1 □

The assignment of $z(t)$ to $b(t)$ defines a map ρ whose domain range and continuity properties we now describe. For later applications we are particularly interested in the continuity properties of the difference $z(t) - b(t)$. Of course continuity of this difference is closely related to the continuity properties of the fixed point of the contraction $J_{\bar{b}}$ as a function of b.

To this end define B to be the space of continuous paths $\gamma \colon [T_0, \infty) \to \mathcal{K}_\Gamma$ so that

$$Sup_{t \in [T_0, \infty)} ||\gamma(t)||_{L^2} e^{\frac{\mu_\Gamma^-}{2}(t-T_0)} < \infty$$

with the norm $||\gamma||_B = Sup_{t \in [T_0, \infty)} ||\gamma(t)||_{L^2} e^{\frac{\mu_\Gamma^-}{2}(t-T_0)}$. Let $\mathcal{P}$ denote the space of C^2-paths $b \colon [T_0, \infty) \to \mathcal{H}_\Gamma$ so that

(1)
$$\ell_{T_0}(b) = \int_{T_0}^{\infty} |\dot{b}| dt < \infty,$$
and

(2)
$$||\dot{b}(t) - \nabla_{b(t)} f_\Gamma||_{L^2} \leq Ce^{-\mu_\Gamma (t-T_0)/2}.$$

Given a path b in $\mathcal{P}$ we define the following functions.

(1) The evaluation $ev \colon \mathcal{P} \to \mathcal{H}_\Gamma$ where $ev(b) = b(0)$.
(2) The length $\ell_{T_0} \colon \mathcal{P} \to R_+$ where $\ell_{T_0}(b) = \int_{T_0}^{\infty} |\dot{b}| dt$.

(3) Define the gradient flow map $\mathcal{F}\colon \mathcal{P} \to B$ by the equation $\mathcal{F}(b) = \dot{b} - \nabla_b f_\Gamma$. (The above definition implies directly that the map $\mathcal{F}$ is well defined.) With this said we may now state the continuity properties of the map ρ.

LEMMA 5.3.4. *For any topology on $\mathcal{P}$ so that the above functions $ev, \ell_{T_0}, \mathcal{F}$ are continuous and with the topology on B induced by $||\cdot||_B$ the map $(Id - \tilde{\Upsilon}_S)\colon \mathcal{P} \to B$ is continuous.*

Proof. The continuity of $ev, \ell_{T_0}, \mathcal{F}$ imply directly that the map $J : \mathcal{P} \times B \to B$ sending (b, γ) to $J_{\overline{b}}(\gamma)$ is continuous. Hence, the fixed point β varies continuously in $\overline{B}$. The desired continuity result follows from this and Inequality 59. □

5.4. L^2-exponential approach to the center manifold

Our next result is used to prove that the path $A(t)$ as in the statement of Theorem 5.2.2 approaches the center manifold exponentially fast in the L^2-norm.

LEMMA 5.4.1. *Fix the notation and hypotheses as in the statement of Theorem 5.2.2. We decompose $A(t) = \Gamma + b(t) + c(t)$ where $b(t) \in \mathcal{H}_\Gamma$ and $c(t) \in H_\Gamma^\perp$. There are constants $C_0 > 0$ and $E_0 > 0$ such that if there is a flat connection $\Gamma + b_\infty \in U_\Gamma$ so that $\sup_{t\in[T_0,\infty)} ||b(t) - b_\infty||_{L^2_2} + ||c(t)||_{L^2_2} = E \leq E_0$, then for all $t \geq T_0$*

$$||c(t)||_{L^2} \leq C_0 E e^{-\mu_\Gamma^-(t-T_0)/2}.$$

Before beginning the proof proper we shall derive the equations that $b(t)$ and $c(t)$ satisfy. To this end we note the following identity. Since $\nabla_{\Gamma+b} f_\Gamma$ is tangent to $\mathcal{H}_\Gamma$ at b we have

$$-*d_\Gamma b + n(b) = \Pi^0(n(b)) + D_{\Pi^0(b)}\zeta(\Pi^0(n(b))). \tag{63}$$

Also since

$$\frac{d(b+c)}{dt} = \nabla_{\Gamma+b+c} f_\Gamma + \wp(b+c, t)$$

we have

$$\frac{db}{dt} = \left(Id + D_{\Pi^0(b)}\zeta\right)\left(\Pi^0(\nabla_{\Gamma+b+c}f_\Gamma + \wp(b+c,t)\right) \tag{64}$$
$$= \left(Id + D_{\Pi^0(b)}\zeta\right)\left(\Pi^0(n(b+c) + \wp(b+c,t)\right).$$

Finally

$$\begin{aligned}\frac{dc}{dt} &= \nabla_{\Gamma+b+c}f_\Gamma + \wp(b+c,t) - \frac{db}{dt} \\ &= -*d_\Gamma b - *d_\Gamma c + n(b+c) + \wp(b+c,t) \\ &\quad - \left(Id + D_{\Pi^0(b)}\zeta\right)\left(\Pi^0(n(b+c) + \wp(b+c,t)\right) \\ &= -*d_\Gamma c + \left(\Pi^\perp - D_{\Pi^0(b)}\zeta \circ \Pi^0\right)\left(n(b+c) - n(b) + \wp(b+c,t))\right)\end{aligned}$$

where to get the last equation we use Equation 63. By Inequality 26 there is a constant $K_0 > 0$ such that

$$||\wp(b+c,t)||_{L^2} \leq K_0 E e^{-\mu t} \tag{65}$$

whenever $b + c \in U_\Gamma$ and $||b - b_\infty||_{L^2_2} + ||c||_{L^2_2} \leq E$.

There is a further decomposition of $c(t)$ as $c^+(t) + c^-(t)$ corresponding to the L^2-orthogonal decomposition along the positive and negative eigenspaces for $-*d_\Gamma$, namely, $H_\Gamma^\perp = H^+ \oplus H^-$. One consequence of Inequality 19 is the following pair of differential inequalities for $c^\pm(t)$.

CLAIM 5.4.2. *There is a constant $E_0 > 0$ such that if $\sup_{t\in[T_0,\infty)} ||b(t) - b_\infty||_{L^2_2} + ||c(t)||_{L^2_2} = E \leq E_0$, then*

$$\frac{d}{dt}||c^+||_{L^2} \geq \mu_\Gamma^+||c^+||_{L^2} - \left(\frac{min\{\mu_\Gamma^-, \mu_\Gamma^+\}}{4}||c||_{L^2} + K_0 E e^{-\mu t}\right), \text{ and} \tag{66}$$

$$\frac{d}{dt}||c^-||_{L^2} \leq -\mu_\Gamma^-||c^-||_{L^2} + \left(\frac{min\{\mu_\Gamma^-, \mu_\Gamma^+\}}{4}||c||_{L^2} + K_0 E e^{-\mu t}\right). \tag{67}$$

Proof. We prove the first inequality leaving the analogous proof of the second inequality to the reader. Let $E_0 = \frac{min\{\mu_\Gamma^-, \mu_\Gamma^+\}}{32 C_1}$ where C_1 is the constant of

Lemma 2.5.3. That lemma implies that if $A(t) = \Gamma + b(t) + c(t)$ satisfies the hypothesis of Lemma 5.4.2 with $E \le E_0$ then

$$||n(b(t)+c(t)) - n(b(t))||_{L^2} \le \frac{min\{\mu_\Gamma^-, \mu_\Gamma^+\}}{4}||c||_{L^2}. \tag{68}$$

Now suppose that t is such that $c^+(t) \ne 0$. Then we have

$$\begin{aligned}
&\frac{1}{2}\frac{d}{dt}(||c^+||^2_{L^2}) \\
&= \; < c^+, \frac{d}{dt}c^+ > \\
&= \; < c^+, \frac{dc}{dt} > \\
&= \; < c^+, -*d_\Gamma c + \left(\Pi^\perp - D_{\Pi^0(b)}\zeta \circ \Pi^0\right)(n(b+c) - n(b) + \wp(b+c,t))) > \\
&\ge \; \mu_\Gamma^+(||c^+||^2_{L^2}) \\
&\quad - ||c^+||_{L^2}|| \left(\Pi^\perp - D_{\Pi^0(b)}\zeta \circ \Pi^0\right)(n(b+c) - n(b) + \wp(b+c,t)))||_{L^2}.
\end{aligned}$$

Using Inequality 58, Inequality 68, and Inequality 65 we have

$$\frac{1}{2}\frac{d}{dt}(||c^+||^2_{L^2}) \;\ge\; \mu_\Gamma^+(||c^+||_{L^2})^2 - \left(\frac{min\{\mu_\Gamma^-, \mu_\Gamma^+\}}{4}||c||_{L^2} + K_0Ee^{-\mu t}\right).$$

The desired inequality follows for all t for which $c^+(t) \ne 0$. By continuity it follows in the closure of this set of t. In the complementary open set where $c^+ = 0$ it is obvious. □

Next we prove a lemma allowing us to estimate the size c^+ in terms of the size of c^-.

CLAIM 5.4.3. *Let E_0 be the constant of Claim 5.4.2. If $\sup_{t\in[T_0,\infty)} ||b(t) - b_\infty||_{L^2_2} + ||c(t)||_{L^2_2} = E \le E_0$, then we have that $||c^+(t)||_{L^2} < ||c^-(t)||_{L^2} + \frac{2K_0}{\mu}Ee^{-\mu t}$.*

Proof. Taking the difference of Inequalities 66 and 67 gives the following differential inequality.

$$\frac{d}{dt}(||c^+||_{L^2} - ||c^-||_{L^2} - \frac{2K_0}{\mu}Ee^{-\mu t}) \ge \frac{min\{\mu_\Gamma^-, \mu_\Gamma^+\}}{2}||c||_{L^2}. \tag{69}$$

Thus $||c^+||_{L^2} - ||c^-||_{L^2} - \frac{2K_0}{\mu}Ee^{-\mu t}$ is a nondecreasing function. If it is negative, then we have established the claim. Assume that it is nonnegative. Then we have

$$\begin{aligned}\frac{d}{dt}||c^+||_{L^2} &\geq \mu_\Gamma^+||c^+||_{L^2} - \frac{min\{\mu_\Gamma^-, \mu_\Gamma^+\}}{4}||c||_{L^2} + K_0Ee^{-\mu t}\\ &\geq \mu_\Gamma^+||c^+||_{L^2} - \frac{min\{\mu_\Gamma^-, \mu_\Gamma^+\}}{4}(2||c^+||_{L^2} - \frac{2K_0}{\mu}e^{-\mu t}) + K_0Ee^{-\mu t}\\ &\geq \frac{\mu_\Gamma^+}{2}||c^+||_{L^2} - \frac{K_0}{2}Ee^{-\mu t}.\end{aligned}$$

Thus, we reach the contradiction that $||c^+(t)||_{L^2}$ is exponentially increasing. □

We are now ready to prove Lemma 5.4.1.

Proof. Let E_0 be the constant of Claim 5.4.2.

$$\begin{aligned}\frac{d}{dt}||c^-||_{L^2} &\leq -\mu_\Gamma^-||c^-||_{L^2} + \frac{min\{\mu_\Gamma^-, \mu_\Gamma^+\}}{4}||c||_{L^2} + K_0Ee^{-\mu t}\\ &\leq -\mu_\Gamma^-||c^-||_{L^2} + \frac{min\{\mu_\Gamma^-, \mu_\Gamma^+\}}{4}(2||c^-||_{L^2} + \frac{2K_0E}{\mu}e^{-\mu(t)})\\ &\quad + K_0Ee^{-\mu(t-T_0)}\\ &\leq -\frac{\mu_\Gamma^-}{2}||c^-||_{L^2} + \frac{\nu min\{\mu_\Gamma^-, \mu_\Gamma^+\}}{16}e^{-\mu t}.\end{aligned}$$

We integrate this inequality to get

$$||c^-(t)||_{L^2} \leq \left(||c_-(T_0)||_{L^2} + \frac{3K_0}{\mu}E\right)e^{-\mu_\Gamma^-(t-T_0)/2} \leq (1 + \frac{3K_0}{\mu})Ee^{-\mu_\Gamma^-(t-T_0)/2}.$$

For a suitable constant C_0 Lemma 5.4.1 follows from this and the previous claim. □

5.5. Proof of Theorem 5.2.2

We are now in a position to finish the proof of Theorem 5.2.2. We begin with

5.5.1. The proof of existence of the map $\tilde{\Upsilon}_S$.. Let $K_1 > 0$ be the constant of Corollary 3.3.2 and let K_2 be the constant of Lemma 3.5.1. Finally let K_3 denote the norm of the restriction $L_3^2([t-\frac{1}{2}, t+\frac{1}{2}] \times N) \to L_2^2(\{t\} \times N)$. Given $\kappa > 0$ choose constants $\epsilon, E_1 > 0$ so that the following conditions hold.

(1) $E_1 \leq E_0$ where E_0 is the constant of Lemma 5.4.1.

(2)

$$E_1 \leq \frac{1}{8K_1C_0}\kappa$$

(3)

$$(\frac{1}{8}C_0\mu_\Gamma^- + 2K_0)E_1 \leq \frac{1}{64\mu_\Gamma^- K_1}$$

(4) ϵ is so small that $\frac{K_1K_3}{2\theta}(9+K_2)\epsilon^\theta < E$ and $2K_1K_2\epsilon \leq \kappa$

Now let ω be an ASD connection satisfying the hypothesis of Theorem 5.2.2 with $\epsilon > 0$ as above. Then according to Theorem 4.3.1 there is flat connection $\Gamma' \in U_\Gamma$ so that for all $t \geq T_0 - 1$ we have $||A(t) - \Gamma'||_{L^2} < \frac{2}{\theta}\epsilon^\theta$. Thus for any $t \geq T_0$ we have

$$||A - \Gamma'||_{L^2([t-1,t+1]\times N)} < \frac{4}{\theta}\epsilon^\theta.$$

Since Γ' is flat we can write $\Gamma' = \Gamma + b_\infty$ with $b_\infty \in \mathcal{H}_\Gamma$. By assumption Lemma 3.3.1 holds for V_Γ. Applying this lemma to Γ' and $A(t) + \alpha(t)dt$ we have for all $t \geq T_0$

$$||\alpha(t)||_{L^2} \leq \frac{1}{8}(\frac{2}{\theta}\epsilon^\theta + K_2\epsilon) \leq \frac{1}{4\theta}(1+K_2)\epsilon^\theta.$$

and hence for any $t \geq T_0$ we have

$$||\alpha||_{L^2([t-1,t+1]\times N)} \leq \frac{1}{2\theta}(1+K_2)\epsilon^\theta.$$

Since K_1 is the constant of Lemma 3.3.2, that result tells us that

$$||\omega - \Gamma'||_{L^2_3([t-\frac{1}{2},t+\frac{1}{2}]\times N)} \leq \frac{K_1}{2\theta}(9+K_2)\epsilon^\theta < E.$$

Thus, for all $t \geq T_0$ we have

$$||A(t) - \Gamma'||_{L^2_{\frac{5}{2}}} \leq \frac{K_1K_3}{2\theta}(9+K_2)\epsilon^\theta < E_1. \tag{70}$$

Let $\Gamma + b(t) + c(t) + \alpha(t)dt$ be the standard decomposition of ω. By the choice of E, Lemma 5.4.1 applies to give

$$||c(t)||_{L^2} < C_0E_1e^{-\mu_\Gamma^-(t-T_0)/2} \leq \frac{1}{8K_1}\kappa. \tag{71}$$

Equation 64 tells us that for all $t \geq T_0$ we have

$$
\begin{aligned}
||\frac{db}{dt} - \nabla_b f_\Gamma||_{L^2} &= ||\left(Id + D_{\Pi^0(b)}\zeta\right)\left(\Pi^0(n(b+c) - n(b) + \wp(b+c,t))\right)||_{L^2} \\
&\leq \frac{\mu_\Gamma^-}{8}||c||_{L^2} + 2K_0E_1e^{-\mu t} \\
&\leq \left(\frac{1}{8}\mu_\Gamma^- C_0E_1 + 2K_0E_1\right)e^{-\mu_\Gamma^-(t-T_0)/2} \\
&\leq \frac{1}{64\mu_\Gamma^- K_1}\kappa.
\end{aligned} \tag{72}
$$

Lemma 5.3.1 now applies to the path $b(t)$. It implies that there is a unique integral curve $B(t) = \rho(b)$ for the flow equation on the center manifold satisfying

$$||b(t) - B(t)||_{L^2} < \kappa/8K_1e^{-\mu_\Gamma^-(t-T_0)/2} \tag{73}$$

for all $t \geq T_0$. We set $\tilde{\Upsilon}_{\mathcal{S}}(\omega) = B(T_0)$. Inequalities 71 and 73 immediately give

$$||A(t) - B(t)||_{L^2} < \kappa/4K_1e^{-\mu_\Gamma^-(t-T_0)/2} \tag{74}$$

for all $t \geq T_0$.

Next we apply Lemma 3.3.1 to the g_0-ASD connection $\tilde{\omega} = B(t) + \Theta(B(t))dt$ and the g-ASD connection $\omega = A(t) + \alpha(t)dt$. By our assumption on g, it follows that for all $t \geq T_0$ we have

$$||\Theta(B(t)) - \alpha(t)||_{L^2} < \frac{1}{8}(||B(t) - A(t)||_{L^2} + K_2\epsilon e^{-\mu t}).$$

Combining this with Inequality 74 gives on for all $t \geq T_0$

$$||\omega - \tilde{\omega}||_{L^2([t-1,t+1]\times N)} \leq (\kappa/2K_1 + K_2\epsilon)e^{-\mu_\Gamma^-(t-T_0)/2} \leq \kappa/K_1e^{-\mu_\Gamma^-(t-T_0)/2}.$$

Now Lemma 3.3.2 implies that for any $t \geq T_0$, we have

$$||\omega - \tilde{\omega}||_{L^2_3([t-\frac{1}{2},t+\frac{1}{2}]\times N} < \kappa e^{-\mu_\Gamma^-(t-T_0)/2},$$

verifying the estimate given in the statement of the theorem. The equivariance of $\tilde{\Upsilon}_{\mathcal{S}}$ under $Stab(\Gamma)$ follows directly from the uniqueness assertion of Lemma 5.3.1.

5.5.2. The proof of continuity of $\tilde{\Upsilon}_{\mathcal{S}}$. To prove that $\tilde{\Upsilon}_{\mathcal{S}}$ is continuous according to Lemma 5.3.4 it suffices to prove that if $\Gamma + b(t) + c(t) + \alpha(t)dt$ is the standard decomposition of $\omega \in Std(\mathcal{S}, g)$ then ev,ℓ_{T_0} and $\mathcal{F}$ are continuous as functions of $b(t)$. The assignment of $A(T_0)$ to A is continuous so the continuity of ev follows directly from the continuity of ζ.

Continuity of ℓ_{T_0}.To prove continuity of ℓ_{T_0} we fix $r > 0$ and find a neighborhood $\mathcal{O} \subset Std(\mathcal{S}, g)$ of ω_1 so that for all $\omega_2 \in \mathcal{O}$ if $\Gamma + b_i(t) + c_i(t) + \alpha_i(t)dt$ are the standard decompositions of ω_i, for $i = 1, 2$, then $|\ell_{T_0}(b_1) - \ell_{T_0}(b_2)| < r$.

To this end fix $\epsilon < min\{\epsilon_0, (\frac{r}{6\theta})^{1/\theta}\}$. There is a neighborhood $\mathcal{O} \subset Std(\mathcal{S}, g)$ of ω_1 so that Lemma 4.7.1 holds for ϵ, $T_1 > T_0$ and $\delta = \frac{r}{6(T_1 - T_0)}$ and all $\omega_2 \in \mathcal{O}$.

Now fix $\omega_2 \in \mathcal{O}$. Then we have

$$\begin{aligned}
|\ell_{T_0}(b_1) - \ell_{T_0}(b_2)| &= \int_{T_0}^{\infty} \left| \left\| \frac{db_1}{dt} \right\|_{L^2} - \left\| \frac{db_2}{dt} \right\|_{L^2} \right| dt \\
&\leq \int_{T_0}^{\infty} \left\| \frac{db_1}{dt} - \frac{db_2}{dt} \right\|_{L^2} dt \\
&\leq \int_{T_0}^{\infty} 2 \left\| \frac{d(A_1 - A_2)}{dt} \right\| dt \\
&\leq 2 \int_{T_0}^{T_1} \left\| \frac{d(A_1 - A_2)}{dt} \right\| dt + 2 \int_{T_1}^{\infty} \left(\left\| \frac{dA_1}{dt} \right\| + \left\| \frac{dA_2}{dt} \right\| \right) dt \\
&\leq 2\delta(T_1 - T_0) + \frac{4}{\theta}\epsilon^{\theta} \leq r/3 + 2r/3 = r
\end{aligned}$$

The Continuity of $\mathcal{F}$.To prove continuity of $\mathcal{F}$ we fix $r > 0$ and find a neighborhood $\mathcal{O}$ of $\omega_1 \in Std(\mathcal{S}, g)$ so that for all $\omega_2 \in \mathcal{O}$ if $\Gamma + b_i(t) + c_i(t) + \alpha(t)dt$ is standard decomposition of ω_i, for $i = 1, 2$, then we have $|\mathcal{F}(b_1) - \mathcal{F}(b_2)|e^{\mu_\Gamma^-(t - T_0)/2} \leq r$ for all $t \geq T_0$.

To begin we need the following generalization of Lemma 5.4.1.

LEMMA 5.5.1. *Fix the notation and hypotheses as in the statement of Theorem 5.2.2 and for $i = 1, 2$ let ω_i be g-ASD connections satisfying the hypotheses of this theorem. Let $E_0 > 0$ be the constant of Lemma 5.4.1. There is a constant*

$C_0 > 0$ so that the following holds. If we write

$$\omega_i = A_i(t) + \alpha_i(t)dt = \Gamma + b_i(t) + c_i(t) + \alpha_i(t)dt$$

where $b_i(t) \in \mathcal{H}_\Gamma$ and $c_i(t) \in H_\Gamma^\perp$. And if there are flat connections $\Gamma + b_{\infty,i}$ with

$$||b_i - b_{\infty_i} + c_i||_{L^2_2} \leq E_0$$

then setting $E = \sup_{t\in[T_0,\infty)} ||A_1(t) - A_2(t)||_{L^2_2}$ we have for all $t \geq T_0$,

$$||c_1(t) - c_2(t)||_{L^2} \leq C_0 E e^{-\mu_\Gamma^-(t-T_0)/2}.$$

The proof of this result is analogous to that of Lemma 5.4.1 and will be given at the end of this subsection.

We will also need the following

LEMMA 5.5.2. *There is a constant $C > 0$ so that for all $b_1, b_2 \in \mathcal{H}'_\Gamma$ we have*

$$||D_{\Pi^o(b_1)}\zeta - D_{\Pi^o(b_2)}\zeta|| \leq C||b_1 - b_2||_{L^2}. \tag{75}$$

Here the norm of the left hand side is the operator norm of the difference as a map from L^2 to L^2.

Proof. This is immediate from the C^2 property of ζ and the commensurablity of the L^2 and riemannian distance in $\mathcal{H}'_\Gamma$. □

Now fix $r > 0$. Choose $E_1 > 0$ so that the following hold.

(1) $E_1 \leq E_0$
(2) $(2(C+2)C_0 + 2K_0 + 4C_1C_0)E_0E_1 < r$ where C is the above constant, C_0 is the constant of Lemma 5.4.1, and C_1 is the constant of Inequality 26.

Let $\mathcal{O} \subset \mathcal{M}_{\ell,w}(\mathcal{S}, g)$ be a neighborhood of ω_1 so that Lemma 4.7.1 holds for $\delta = \frac{E_1}{K_3}$, some $T_1 > T_0$ and $\epsilon = \epsilon_0$ and all $\omega_2 \in \mathcal{O}$. Let ω_1 and ω_2 be as above. Then according to Equation 64 we have

$$
\begin{aligned}
&||\mathcal{F}(b_1) - \mathcal{F}(b_2)||_{L^2} \\
= \quad & || \left(Id + D_{\Pi^0(b_1)}\zeta\right) \left(\Pi^0(n(b_1 + c_1) - n(b_1) + \wp(b_1 + c_1, t)\right) \\
& - \left(Id + D_{\Pi^0(b_2)}\zeta\right) \left(\Pi^0(n(b_2 + c_2) - n(b_2) + \wp(b_2 + c_2, t)\right)||_{L^2} \\
& (\text{ By Inequality 75}) \\
\leq \quad & C||b_1 - b_2||||(\Pi^0(n(b_1 + c_1) - n(b_1) + \wp(b_1 + c_1, t)) \\
& +2||(\Pi^0(n(b_1 + c_1) - n(b_1) + \wp(b_1 + c_1, t)) \\
& -(\Pi^0(n(b_2 + c_2) - n(b_2) + \wp(b_2 + c_2, t))||_{L^2} \\
& (\text{ By Inequality 58}) \\
\leq \quad & C||b_1 - b_2||_{L^2}||(\Pi^0(n(b_1 + c_1) - n(b_1) + \wp(b_1 + c_1, t))||_{L^2} \\
& +2||n(b_1 + c_2) - n(b_2 + c_2) - n(b_1) + n(b_2)||_{L^2} \\
& +2||n(b_1 + c_1) - n(b_1 + c_2)||_{L^2} \\
& +2||\wp(b_1 + c_1, t) - \wp(b_2 + c_2, t)||_{L^2} \\
& (\text{ By Inequality 19}) \\
\leq \quad & (C + 2)C_1||b_1 - b_2||_{L^2}(||c_1||_{L^2}||c_2||_{L^2} + K_0 e^{-\mu t}||b_1 + c_1||_{L^2}) \\
\leq \quad & +2C_1||b_1||_{L^2}||c_1 - c_2||_{L^2} + 2K_0 e^{-\mu t}||b_1 + c_1 - (b_2 + c_2)||_{L^2} \\
& (\text{ By Lemma 5.4.1 }) \\
\leq \quad & 2(C + 2)E_1(C_0 E_0 e^{-\mu_{\Gamma}^{-}(t - T_0)/2} + K_0 E_0 e^{-\mu t}) + 4C_1 E_0 C_0 E_1 e^{-\mu_{\Gamma}^{-}(t - T_0)/2} \\
& (\text{ and Lemma 5.5.1 }) \\
& +2K_0 e^{-\mu t} E_1 \\
\leq \quad & r e^{-\mu_{\Gamma}^{-}(t - T_0)/2}
\end{aligned}
$$

This finishes the proof of the continuity of $\mathcal{F}$, and hence completes the proof of the continuity of $\tilde{\Upsilon}$ modulo the proof of Lemma 5.5.1.

We now return to the proof of Lemma 5.5.1.

Proof. The result will follow from the proof of Lemma 5.4.1 if we can demonstrate analogous inequalities to those of Claim 5.4.2 for $c_1 - c_2$. Let us now do so.

CLAIM 5.5.3. *With the notation and hypotheses as in Lemma* 5.5.1 *there is a constant* $C_1 > 0$ *so that if* $\sup_{t\in[T_0,\infty)} ||A_1(t) - A_2(t)||_{L^2_2} = E$, *then*

$$\frac{d}{dt}||(c_1 - c_2)^+||_{L^2} \geq$$
$$\mu_\Gamma^+||(c_1 - c_2)^+||_{L^2} - \left(\frac{min\{\mu_\Gamma^-, \mu_\Gamma^+\}}{4}||(c_1 - c_2)||_{L^2} + 2K_0Ee^{-\mu t}\right),$$

and

$$\frac{d}{dt}||(c_1 - c_2)^-||_{L^2} \leq$$
$$-\mu_\Gamma^-||(c_1 - c_2)^-||_{L^2} + \left(\frac{min\{\mu_\Gamma^-, \mu_\Gamma^+\}}{4}||c_1 - c_2||_{L^2} + 2K_0Ee^{-\mu t}\right).$$

Proof. As in the proof of Claim 5.4.2 we have.

$$\frac{1}{2}\frac{d}{dt}(||(c_1 - c_2)||^2_{L^2}) \geq$$
$$\mu_\Gamma^+(||(c_1 - c_2)^+||^2_{L^2})$$
$$-||(c_1 - c_2)^+||_{L^2}|| \left(\Pi^\perp - D_{\Pi^0(b_1)}\zeta \circ \Pi^0\right)(n(b_1 + c_1) - n(b_1) + \wp(b_1 + c_1, t)))$$
$$\left(\Pi^\perp - D_{\Pi^0(b_2)}\zeta \circ \Pi^0\right)(n(b_2 + c_2) - n(b_2) + \wp(b_2 + c_2, t)))||_{L^2}$$

The desired inequality follows from this as before using Inequality 58 and Inequality 19.

□

5.6. Weighted Sobolev space continuity of $\omega - \tilde{\omega}$.

To finish the proof of Theorem 5.2.2 we must verify the continuity assertion in the last paragraph of the statement of that theorem.

What we must show is that given $r > 0$ and $\omega_1 \in Std(\mathcal{S}, g)$ there is a neighborhood $\mathcal{O}$ of ω_1 so that for all $\omega_2 \in \mathcal{O}$ if $\tilde{\omega}_i$, for $i = 1, 2$ denote the corresponding g_0

ASD connections obtained from Theorem 5.2.2 then

$$||(\omega_1 - \tilde{\omega}_1) - (\omega_2 - \tilde{\omega}_2)||_{L^2_{2,\delta}([T_0,\infty)\times N)} < r.$$

Lemma 3.5.2 immediately implies that it suffices to prove the L^2_δ-version of the above inequality:

(76) $$||(\omega_1 - \tilde{\omega}_1) - (\omega_2 - \tilde{\omega}_2)||_{L^2_\delta([T_0,\infty)\times N)} < r.$$

Let $\omega_i = \Gamma + b_i(t) + c_i(t) + \alpha_i(t)dt$ and $\tilde{\omega}_i = \Gamma + \tilde{b}_i(t) + \tilde{\alpha}_i(t)dt$, for $i = 1, 2$ be the standard decompositions then we have

$$\begin{aligned}(\omega_1 - \tilde{\omega}_1) - (\omega_2 - \tilde{\omega}_2) &= ((b_1(t) - \tilde{b}_1(t)) - (b_2(t) - \tilde{b}_2(t))) + c_1(t) - c_2(t) \\ &\quad +((\alpha_1(t) - \tilde{\alpha}_1(t)) - (\alpha_2(t) - \tilde{\alpha}_2(t)).\end{aligned}$$

The functions induced by $\ell_{T_0}, ev, \mathcal{F}$ on $Std(\mathcal{S}, g)$ were proved to be continuous in the previous section. Thus Lemma 5.3.4 implies that the map $b(t) \to b(t) - \tilde{b}(t)$ is continuous with respect to the norm $||\cdot||_B$. Thus, we can find a neighborhood $\mathcal{O}_1$ of ω_1 so that for all $\omega_2 \in \mathcal{O}_1$ we have for all $t > T_0$

$$||(b_1(t) - \tilde{b}_1(t)) - (b_2(t) - \tilde{b}_2(t))||_{L^2(\{t\}\times N)} \le \frac{r(\mu_\Gamma^- - \delta)}{8} e^{-\mu_\Gamma^-(t-T_0)}.$$

According to Lemma 3.3.1 we have

$$||((\alpha_1(t) - \tilde{\alpha}_1(t)) - (\alpha_2(t) - \tilde{\alpha}_2(t))||_{L^2} \le$$

(77) $$\frac{1}{8}||c_1(t) - c_2(t)||_{L^2} + ||h||_{C^1}||F_{\Gamma+b_1(t)+c_1(t)} - F_{\Gamma+b_2(t)+c_2(t)}||_{L^2}.$$

Find $E_1 > 0$ so that Lemma 5.5.1 holds for $\nu = \frac{r(\mu_\Gamma^- - \delta)}{4}$. Also find E_2 so that the Sobolev embedding theorems imply that if

$$||(b_1(t) + c_1(t)) - (b_2(t) + c_2(t))||_{L^2_3([T-1,T+1]\times N)} \le E_2.$$

then

$$||F_{\Gamma+b_1(t)+c_1(t)} - F_{\Gamma+b_2(t)+c_2(t)}||^2_{L^2(\{t\}\times N)} \le \frac{r(\mu - \delta)}{16}.$$

The neighborhood $\mathcal{O}$ is determined by the following requirements.

(1) $\mathcal{O} \subset \mathcal{O}_1$.

(2) Lemma 4.7.1 holds for $\delta = \frac{1}{K_3} min(E_1, E_2)$ and $\epsilon = \frac{r(\mu-\delta)}{8}$ and some $T_1 > T_0$.

(3) On the compact set $[T_0, T_1] \times N$ we have that

$$||(\omega_1 - \tilde{\omega}_1) - (\omega_2 - \tilde{\omega}_2)||_{L^2_\delta([T_0,\infty)\times N)} < r/2.$$

The assumptions on $\mathcal{O}$ and Inequality 77 imply the following inequalities.

$$\begin{aligned}
&\int_{T_0}^{\infty} ||(\omega_1 - \tilde{\omega}_1) - (\omega_2 - \tilde{\omega}_2)||^2_{L^2(\{t\}\times N)} e^{\delta t} \\
\leq\ & r/2 + \int_{T_1}^{\infty} ||(b_1(t) - \tilde{b}_1(t)) - (b_2(t) - \tilde{b}_2(t))||_{L^2(\{t\}\times N)} e^{\delta t} \\
& +2 \int_{T_1}^{\infty} ||c_1(t) - c_2(t)||^2_{L^2(\{t\}\times N)} e^{\delta t} \\
& + \int_{T_1}^{\infty} ||h||^2_{C^2} ||F_{\Gamma + b_1(t)+c_1(t)} - F_{\Gamma+b_2(t)+c_2(t)}||^2_{L^2} e^{\delta t} \\
\leq\ & r/2 \int_{T_0}^{\infty} (3\frac{r(\mu_\Gamma^- - \delta)}{16} e^{(\delta - \mu_\Gamma^-)(t-T_0)} + 2\frac{r(\mu - \delta)}{8} e^{(\delta-\mu)t}) \\
\leq\ & r.
\end{aligned}$$

This completes the proof of Theorem 5.2.2. □

CHAPTER 6

Geometric limits as we stretch the tube out

Let M be a closed oriented manifold and $\overline{N} \hookrightarrow M$ an closed orientable codimension one submanifold. In order to study the Donaldson polynomial invariants of M we shall 'pull the metric apart' along $\overline{N}$ and study the limiting behavior of the moduli space in terms of the moduli spaces of the components of $M \setminus \overline{N}$ and gluing data along $\overline{N}$. In practice, it is important to be able to understand when energy is 'lost' in the tube as the manifold is pulled apart. The notion of geometric limits seems particularly well-suited to analyzing this phenomenon. Before we get into the specifics of our particular case, we give a general discussion on this notion.

6.1. General geometric limits

Suppose that we have a sequence of complete riemannian manifolds (M_n, g_n) with base points $p_n \in M_n$. We say that a riemannian manifold (X, g) with base point b is *the geometric limit of the sequence* (M_n, g_n, p_n) if:

(1) (X, g) is complete and connected,
(2) for every $t \geq 0$ and all n sufficiently large there are compact submanifolds with boundary $A \subset X$ and $B_n \subset M_n$ containing the balls of radius t centered at b and p_n respectively and a diffeomorphism ϕ_n from (A, b) to (B_n, p_n) so that the metrics $\phi_n^*(g_n|B_n)$ converge in the C^∞ topology to $g|A$.

More generally, suppose that in each M_n we have a subset $P_n = \{p_n^1, \ldots, p_n^k\}$ of base points. We assume that for all $i \neq j$ $lim_{n\mapsto\infty} d(p_n^i, p_n^j) = \infty$. A riemannian manifold (X, g) with a subset $B = \{b^1, \ldots, b^k\}$ is *the geometric limit of the sequence* (M_n, g_n, P_n) if:

(1) X has k components $X_1, \ldots, X_k$ with $b^i \in X_i$, and
(2) for each $i \leq k$ we have $(X_i, g|X_i, b^i)$ is *the geometric limit of the sequence* (M_n, g_n, p_n^i).

Suppose that (X, g, b) is the geometric limit of a sequence (M_n, g_n, p_n). Suppose that for each n we have a bundle $E_n \to M_n$ and a connection ω_n on E_n. We say that a connection ω on a bundle $E' \to X$ is the geometric limit of the ω_n if for every $t \geq 0$ and all n sufficiently large there are compact submanifolds with boundary $A \subset X$ and $B_n \subset M_n$ containing the balls of radius t centered at b and p_n respectively and diffeomorphisms ϕ_n from (A, b) to (B_n, p_n) covered by a bundle maps $\tilde{\phi}_n : E'|A \to E|B_n$ so that:

(1) $\phi_n^*(g_n|B_n)$ converges in the C^∞-topology to $g|A$ and
(2) $lim_{n\mapsto\infty} \tilde{\phi}_n^*(\omega_n|B_n) = \omega|A$ in the $L_2^2(A)$-topology.

As above there is an analogous notion of the geometric limit of a sequence of connections ω_n over riemannian manifold (M_n, g_n) with a finite set of base points $P_n = \{p_n^1, \ldots, p_n^k\}$.

We also need the notion of geometric limits which are generalized connections. Let (M_n, g_n, p_n) be as above with geometric limit (X, g, b). For each n let ω_n be a connection on a bundle $E_n \to M_n$. We say that a connection ω on a bundle $E' \to X$ and a finite subset of distinct points $\{x_1, \ldots, x_s\}$ in X with multiplicities $m_1, \ldots, m_s$ is the *geometric limit* of the ω_n provided that for every $t > 0$ and all n sufficiently large there are diffeomorphisms $\phi_n : A \to B_n$ as before and bundle maps $\tilde{\phi}_n$ covering $\phi_n|(A \setminus \{x_1, \ldots, x_s\})$ such that

(1) $\tilde{\phi}_n^*\omega_n$ converges in the $L^2_{2,loc}$ topology to $\omega|(A \setminus \{x_1, \ldots, x_s\}$, and

(2)

$$lim_{n\mapsto\infty}\phi_n^*(-trF_{\omega_n}\wedge F_{\omega_n}) = -trF_\omega \wedge F_\omega + \sum_{i=1}^{s} 8\pi^2 c(G)n_i\delta_{x_i}$$

in the weak topology on the space of measures on A.

Here $c(G)$ is the ratio $\frac{B_G}{B_{SU(2)}}$ of Chapter 8 of [1] or equivalently $4c(G)$ is the smallest positive integer that is the first Pontraygin class of the adjoint bundle of a principal G-bundle over the four sphere. In particular $c(SU(2)) = 1$.

We call such a pair $(\omega, \{(x_1, m_1), \dots, (x_j, m_j)\}$ *generalized connection*. Its *energy* is the sum of the energy of the connection and $\sum_{i=1}^{j} 8\pi^2 c(G)m_i$. Notice that the energy of a geometric limit of connections is less than or equal to the liminf of the energies of the connections in the sequence. A sequence of connections is said to *converge without loss of energy* to an idealized connection over X if it has a geometric limit and the energy of this limit is equal to the limit of the energies of the connections in the sequence.

6.1.1. Uhlenbeck's compactness theorem. The following is the usual Uhlenbeck limit theorem in this context.

THEOREM 6.1.1. *Let (M_n, g_n, P_n) be a sequence of riemannian 4-manifolds with base points. Suppose that this sequence has a geometric limit (X, g, B). Let $E_n \to M_n$ be principal G-bundles. Let $\{\omega_n\}$ be connections on E_n so that for all n, ω_n is ASD with respect to the metric g_n and the energy of ω_n is independent of n. Then there is a generalized g-ASD L^2-connection on X which is the geometric limit of a subsequence of $\{\omega_n\}$. Its energy is at most that of the ω_n.*

6.2. Stretching the neck

Next we discuss a family of metrics exhibiting the sort of geometric limits we will study. Let $\overline{N} \subset M$ be as above. Fix an identification of $[-2, 2] \times \overline{N}$ with a neighborhood $\nu_{\overline{N}}$ of $\overline{N}$ in M. Fix a metric g on M which restricts to the product metric $ds^2 + g_{\overline{N}}$ on $[-2,2] \times \overline{N}$. For each component $\overline{C}$ of $\overline{N}$ consider a smooth function $\psi_{\overline{C}}(s)$ defined on $[-2,2]$ satisfying:

(1) $\psi_{\overline{C}}(s) = |1|$ near $s = 2, -2$,

(2) $\psi_{\overline{C}}(s) > 0$ for all s,

We define a riemannian metric g' on M which agrees with g on $M - \nu_{\overline{N}}$ and which are given by

$$g' = \psi_{\overline{C}}(s)^2 ds^2 + g_{\overline{C}}$$

in $\nu_{\overline{C}}$. We call

$$L(\overline{C}) = \int_{-2}^{2} \psi_{\overline{C}}(s) ds$$

the length of the tube across $\overline{C}$ *in* (M, g'). Up to unique isometry which is the identity on $M \setminus \nu_{\overline{N}}$ the tube lengths $L(\overline{C})$ are the only invariants of g'. We call g' *a metric of tube lengths* $L(\overline{C})$ constructed from g.

Suppose that we have a sequence of metrics g_n of tube lengths $L_n(\overline{C}) \mapsto \infty$ as $n \mapsto \infty$. With respect to a base point in each component of $M \setminus \nu_{\overline{N}}$ this sequence has a geometric limit which is identified with

$$X = \left(M \setminus \nu_{\overline{N}}\right) \cup \left([0,\infty) \times N\right)$$

with the metric g_∞ which agrees with g on $M \setminus \nu_{\overline{N}}$ and is a product metric $dt^2 + g_N$ on the infinite cylinder. Here, N is the orientation double cover of $\overline{N}$ and g_N is the metric induced from $g_{\overline{N}}$. We call (X, g_∞) *the cylindrical-end manifold associated to* $(M, \overline{N}, g)$.

There is an involution of N given by change of orientation. The manifold N is canonically identified with $\partial\nu_{\overline{N}}$ so that the end of X is modeled on $[0,\infty) \times N$. We choose this identification so that the image of $\{0\} \times N$ is $\partial\nu_{\overline{N}} \subset X$. Let $\tau \colon X \to [-1,\infty)$ be a smooth function which agrees with the first coordinate on $[0,\infty) \times N$ and such that $\tau \leq 0$ on the rest of X. For $T \geq 0$ define $N_T = \tau^{-1}(T)$. Let C be a component of N. C_T^+ is the component of N_T corresponding to C and C_T^- is the component of N_T corresponding to $-C$.

In this context Theorem 4.0.1 gives a continuous map $\partial\colon \mathcal{M}_w(X,g) \to \chi(N)$. Notice the natural involution on N gives rise to a natural involution on $\chi(N)$. We say a connection ω has *compatible boundary values* if $\partial([\omega])$ is invariant under this involution. There is an analogous notion in the based context.

6.3. Convergence without loss of energy on long tubes

6.3.1. Compatible boundary values. In order to prove that the open sets that we will describe in later sections cover the entire moduli space we must study the limiting behavior as the tube length goes to infinity of ASD connections on a principal bundle over M. Of course, Uhlenbeck's compactness theorem applies in this context to give geometric limit connections on the cylindrical manifold As we noticed above, in taking these geometric limits there can be a loss of energy. In the case where where there is no loss of energy we can make a stronger statement by appealing to Simon's length estimate.

PROPOSITION 6.3.1. *Let $\{g_n\}_{n=1}^{\infty}$ be a sequence of metrics on M with tube lengths $L_n(\overline{C})$ with $\lim_{n\to\infty} L_n(\overline{C}) = \infty$. Let $\{\omega_n\}$ be a sequence of connections on E so that for all n, ω_n is ASD with respect to metric g_n. Suppose that $\{\omega_n\}$ has a geometric limit $(\omega, \{(x_1, m_1), \ldots, (x_j, m_j)\})$, and suppose that there is no loss of energy in this limit. Then the connection ω has compatible boundary values.*

Proof. Fix a component C of N. Let Γ_+ denote the restriction of the limiting connection ω to the end of X corresponding to C and Γ_- the limiting connection restricted to the end corresponding to $-C$. We show that for any neighborhood U of Γ_+ in $\mathcal{B}(C)$ and for all T sufficiently large the restriction of ω to C_T^- is gauge equivalent into U. There is no loss in generality in assuming we have an ϵ_0 Simon patch pair $V \subset U$ centered at Γ_+ and that for T sufficiently large $\omega|[T,\infty) \times C^+$ is in standard form with respect to U. Thus for all T sufficiently large and given

T for all i sufficiently large the restriction of ω_i to C_T is gauge equivalent into V. The convergence without loss of energy implies the following.

CLAIM 6.3.2. *For all T sufficiently large and for all i sufficiently large the energy of ω_i between C_T^+ and C_T^- is less than ϵ_0.*

For T and i so large that the claim holds we can apply Theorem 4.3.1 to conclude that restriction of ω_i to C_T^- is gauge equivalent into U. The result now follows from the geometric convergence of ω_i to ω. (Notice that there is no x_i in $[T,\infty)\times C$ nor in $[T,\infty)\times(-C)$.) □

We need a result that produces convergence without loss of energy for any sequence of ASD connections. In order to get this convergence one must take a geometric limit that has extra components. This is analogous to the bubbling-off phenomenon in the usual Uhlenbeck limits. Each of these extra components will be a copy of $R\times C$ where C is a component of $\overline{N}$. The number of such copies will depend on the sequence under investigation.

6.3.2. The effect of extra base points. Let us fix a sequence of metrics g_n on M of tube lengths $L_n(\overline{C})$ where $L_n(\overline{C})\mapsto\infty$ as $n\mapsto\infty$. Suppose that we use as a base set P_n for M_n a subset consisting of one point in each component of $M-\nu_{\overline{N}}$ and for each component $\overline{C}$ of $\overline{N}$ we take $r_{\overline{C}}\geq 0$ points $\{c_n^i\}_{i=1}^{r_{\overline{C}}}$ in $[-2,2]\times\overline{C}$. We choose the c_n^i of the form (t_n^i,c) with $t_n^1<t_n^2<\cdots<t_n^{r_{\overline{C}}}$ for $c\in\overline{C}$ a fixed base point. We call the c_n^i *extra base points.* (Of course, we must arrange that the distance in (M,g_n) between any pair of the base goes to infinity as $n\mapsto\infty$.) The geometric limit of (M,g_n,P_n) is isometric to the disjoint union of the geometric limit of (M,g_n), and for each $\overline{C}$, $r_{\overline{C}}$ copies of $R\times\overline{C}$ with base point $(0,c)$. It is called *the geometric limit of (M,g_n) with the extra base points.* The components of the form $R\times\overline{C}$ are indexed by $i=1,\cdots r_{\overline{C}}$ in the sense that the component indexed by i has a base point corresponding to (t_n^i,c) in (M,g_n).

There is another way to view this process of adding extra base points. The geometric limit of (M, g_n) with extra base points is also the geometric limit of a sequence of metrics with tube lengths going to infinity constructed using a codimension-1 submanifold $\overline{N}' \subset M$ where each component of $\overline{N}'$ is a parallel copy of some component of $\overline{N}$. Because of this description, the ends of Y are identified with N' and hence there is a notion of compatible boundary values for a connection on Y.

6.3.3. The main compactness theorem. Here is our main result about convergence without loss of energy and with compatible boundary values.

THEOREM 6.3.3. *Let (M, g) and $\overline{N}$ be as above. Let g_n be a sequence of metrics on M associated to g with tube lengths going to infinity. Let $\{\omega_n\}$ be a sequence of connections on E so that for all n, ω_n is ASD with respect to the metric g_n. Assume that all the ω_n have the same energy, say $2\pi^2\ell$. Then, after replacing the sequence by a subsequence, there exists a sequence of extra base points $c_n^i \in [-2,2] \times \overline{N}$ such that the following holds. Let Y be the geometric limit of (M, g_n) with the extra base points. Then the sequence ω_n converges without loss of energy to an idealized connection over Y and this connection has compatible boundary values.*

Proof. Fix $\epsilon_0 > 0$ as in Theorem 4.3.1 for $\overline{N}$. By a *tube* in M we mean a submanifold $T \subset ([-2,2] \times \overline{N})$ of the form $[a,b] \times \overline{C}$ where $\overline{C}$ is a component of $\overline{N}$. The length of this tube in (M, g_n) is simply the g_n-distance between its ends. After passing to a subsequence, for each n we can find a collection of disjoint tubes $T_n^1, \ldots, T_n^r$ in $\nu_{\overline{N}}$ such that:

(1) for each i the length of T_n^i goes to infinity as $n \mapsto \infty$,
(2) for all i and n we have $\int_{T_n^i} (|F_{\omega_n}|_{g_n})^2 < \epsilon_0$, and
(3) there is a constant $C > 0$ such that for any tube $T \subset M - \cup_i T_n^i$ with $\int_T (|F_{\omega_n}|_{g_n})^2 < \epsilon_0$ has length in (M, g_n) less that C.

Since the total energy of the ω_n is bounded independent of n, it follows that the diameter of each component of $M_n - \cup_i T_n^i$ is bounded independent of n. The set

P_n of base points for (M, g_n) associated to the sequence ω_n is constructed by so that there is one point in each component of $M - \cup_i T_n^i$.

We form the geometric limit (Y, g, B) of the sequence (M, g_n, P_n). By Uhlenbeck's compactness theorem, after passing to a subsequence we can assume that the ω_n converge geometrically to a generalized g-ASD connection on Y. The fact that there is no loss of energy in this limit is a direct consequence of the following claim.

CLAIM 6.3.4. *Fix a sequence of ASD connections θ_n on tubes $[-n, n] \times \overline{C}$ of energy less that ϵ_0. Choose a base point $p_n \in [-n, n] \times \overline{C}$ so that the distant from p_n to the boundary of the tube goes to infinity as $n \mapsto \infty$. Suppose that with respect to this sequence of base points the sequence of connections has a geometric limit. Then that limit is a flat connection on $(-\infty, \infty) \times \overline{C}$.*

Proof. By Theorem 4.3.1, after passing to a subsequence we can assume that each θ_n is in standard form with respect to an ϵ_0-Simon patch pair $V_\Gamma \subset U_\Gamma$. Thus, the limiting connection θ is an ASD connection on the infinite cylinder which is also in standard form with respect to $V_\Gamma \subset U_\Gamma$. Hence, its boundary values are flat connections in U_Γ. These connections have the same Chern-Simons invariant. By Equation 2 this implies that the energy of θ is zero; i.e., θ is flat. □

The result about compatible boundary values follows exactly as in the proof of Proposition 6.3.1. □

DEFINITION 6.3.5. *Let us define $\Delta(CS)$ to be the minimum over all pairs of flat connections Γ and Γ' on $\overline{\eta}$ on which the Chern-Simons function takes distinct values of*

$$|CS(\Gamma) - CS(\Gamma')|.$$

Notice that $\Delta(CS)$ is an invariant of $\overline{\eta}$ independent of the connection we have chosen as the reference connection in defining CS.

COROLLARY 6.3.6. *Let (M, g) and $\overline{N}$ be as above. Let g_n be a sequence of metrics on M associated to g with tube lengths going to infinity. Let (X, g_∞) be the cylindrical-end manifold associated to $(M, \overline{N}, g)$. Let $\{\omega_n\}$ be a sequence of connections on E so that for all n, ω_n is ASD with respect to the metric g_n. Assume that all the ω_n have the same energy, say $2\pi^2\ell$. Suppose that the sequence ω_n has a geometric limit ω on X and suppose that the boundary values of ω are not compatible. Then the energy of ω is at most*

$$2\pi^2\ell - \Delta(CS).$$

Part II

The Thickened moduli space

In this part we shall introduce a local thickening of the L^2-moduli space centered at any flat connection on η. In Theorem 7.4.3 we show that the thickening centered at Γ maps to the center manifold $\mathcal{H}_\Gamma$ with the preimage of the stable set of the critical set being identified with an open subset of the L^2-moduli space. In the second chapter of this part we show how the fibers of the map from the thickened moduli space to the center manifold are identified with δ-decaying moduli spaces. This allows us to study their formal deformation theory by the usual techniques. In Theorem 9.0.1 we show that in the case when the gauge group is $SU(2)$ or $SO(3)$, then for a generic metric these thickened moduli spaces are smooth manifolds and the map to the center manifold is a generic map. In the last (and somewhat technical) chapter, we prove the requisite slice result which is needed for the generic metrics theorem.

CHAPTER 7

The definition of the thickened moduli space

In this part, as in the first part, we fix a manifold X with cylindrical end modeled on $[0,\infty)\times N$, a compact, simple Lie group G, a principal G-bundle $E\to X$ and a metric $g\in B$. We set $w=w(E)\in H^2(X;\pi_1(G))$, the characteristic class of E.

In the last chapter we found an open covering $\left\{\mathcal{M}^0_{\ell,w}(\mathcal{S}_\Gamma,g)\right\}$ of the based moduli space $\mathcal{M}^0_{\ell,w}(X,g)$ of ASD connections on E with energy $8\pi^2\ell$. This open cover is indexed by Simon data $\mathcal{S}_\Gamma=(V_\Gamma,T_0)$. We also gave projection maps

$$\Upsilon_{\mathcal{S}_\Gamma}:\mathcal{M}^0_{\ell,w}(\mathcal{S}_\Gamma,g)\to Y_\Gamma\times_{Stab(\Gamma)}\eta_{(n)}$$

where $Y_\Gamma\subset\mathcal{H}_\Gamma$ is the stable set for the restriction, Ξ_Γ, of the vector field ∇f_Γ to the center manifold $\mathcal{H}_\Gamma\subset\mathcal{K}_\Gamma$. Each $h\in Y_\Gamma$ determines a flow line $h(t)$ for Ξ_Γ on the interval $[T_0,\infty)$. This flow line yields a finite energy g_0-ASD connection $\tilde\omega_h=h(t)+\Theta(h(t))$ defined on $[T_0,\infty)\times\eta$. We showed that for every $h\in Y_\Gamma$ any $x\in\Upsilon^{-1}_{\mathcal{S}_\Gamma}([h,f])$ has a representative pair $[\omega,f]$ where $\omega|([T_0,\infty)\times\eta=A(t)+\alpha(t)dt$ is in standard form with respect to Γ and $A(t)$ approaches $h(t)$ exponentially fast. Our purpose is to use these results to provide a context in which one can analyze the local structure of the L^2 moduli space.

Of course, the stable set is contained in the center manifold. This potentially gives the stable set more structure. (For example, it is possible that the stable set is Whitney stratified.) One might hope that the L^2 moduli space is contained in an analogous way inside a more regular object. We will show that this is the case locally by constructing the *thickened moduli space*. The thickened moduli space has another useful feature, one that we do not exploit here but which is extremely important in other contexts: Namely, it provides the model space required for a general gluing theorem.

The idea for thickening the moduli space is simple. We would like consider the ASD connections corresponding to flow lines with arbitrary initial conditions in the center manifold. We cannot in general do this as the flow on the center manifold need not be complete. To get around this problem we use a cutoff function to obtain a truncated version Ξ_Γ^{tr} of Ξ_Γ. We then consider a space of connections which decay exponentially to a connection corresponding to a flow line of Ξ_Γ^{tr}. Of course we need to modify the ASD equation to get an equation compatible with the truncated vector field.

This chapter begins with a discussion of the weighted Sobolev spaces that will be used in the construction. Then we introduce the space of connections which differ on the end of X from the g_0-ASD connection corresponding to a flow line for Ξ_Γ^{tr} by a $L^2_{2,\delta}$ one-form. We also introduce the appropriate gauge group for this context and form the quotient space. Next we consider the space of $L^2_{2,loc}$ connections on E and its quotient by the group of $L^2_{3,loc}$ gauge transformations. Then we prove that that there is a well-defined continuous inclusion of these quotient spaces. Although the topology of the domain is finer than that of the image, when we consider the subspaces of g-ASD connections these topologies agree. Thus the L^2 moduli space has an open cover which can be analyzed in terms of the vector field Ξ_Γ and the weighted Sobolev spaces.

The last section of this chapter introduces a completion of the thickened moduli space.

7.1. δ-decaying forms

Let $\delta > 0$ be given. Fix a product structure $\varphi\colon [0,\infty) \times \eta \to E|([0,\infty) \times N)$, which we consider as an identification. We also fix a C^∞-function $\tau\colon X \to [0,\infty)$ with the following properties:

(i) On $[1,\infty) \times N \subset X$ the function τ agrees with the natural projection onto $[1,\infty)$.

(ii) $\tau\Big|\Big(X - ([0,\infty) \times N)\Big) = 0$.

(iii) $\tau\big([0,1] \times N\big) \subset [0,1]$.

Using the function τ we can define the spaces of L^2_δ-forms with values in adE. Define the L^2_δ-inner product of C^∞ forms α, β with compact support to be

$$< \alpha, \beta >_{L^2_\delta} = \int_X e^{\tau\delta} tr(\alpha \wedge *\beta)$$

where $*$ is the usual Hodge $*$-operator with respect to the metric g_0 on X. We denote the completion of the subspace of $\Omega^i(X; adE)$ of compactly supported forms with respect to the L^2_δ-inner product by $\Omega^i_\delta(X; adE)$, or $\Omega^i_\delta(X)$ when the bundle is clear from the context.

If A^* denotes the formal L^2-adjoint for a differential operator A on $\Omega^*(X; adE)$, then the formal adjoint of A with respect to the L^2_δ-inner product is $e^{-\tau\delta}A^*e^{\tau\delta}$.

Now let us extend this inner product to an $L^2_{k,\delta}$-inner product. To do this we fix a C^∞-connection ω on E with the property that ω on $E|([0,\infty) \times N)$ is the pullback via φ of a connection on η. Let

$$\nabla\colon \Omega^i(X; adE \otimes \Lambda^j T^*X) \to \Omega^i(X; adE \otimes \Lambda^j T^*X \otimes T^*X)$$

denote the covariant derivative of this connection and the metric on X. For each integer $k \geq 0$ and each compactly supported i-form α on X we define

$$\left(\|\alpha\|_{L^2_{k,\delta}}\right)^2 = \int_X e^{\tau\delta}\left(|\alpha|^2 + |\nabla(\alpha)|^2 + \cdots + |\nabla^k(\alpha)|^2\right). \tag{78}$$

We denote the completion of this normed space by $\Omega^i_{k,\delta}(X; adE)$ or $\Omega^i_{k,\delta}(X)$ if E is clear from the context.

Notice that the norm depends on the choice of the connection, but if ω_0 and ω_1 are two C^∞-connections both of whose restrictions to $E|([0,\infty) \times N$ are pullbacks via φ of connections on η then the norms defined by ω_0 and ω_1 are commensurate. In particular, the topologies they define are the same.

Let us recall the weighted version of the Sobolev multiplication theorems which will be needed in what follows.

LEMMA 7.1.1. *There is a constant C depending on the Riemannian manifold X such that for any $1 \leq k, k' \leq 3$ and any $0 \leq r \leq min(k,k')$ if $\alpha \in \Omega^i_{k,\delta}(X)$ and $\beta \in \Omega^{i'}_{k',\delta}(X)$ then $\alpha \wedge \beta \in \Omega^{i+i'}_{r,2\delta}(X)$ and*

$$||\alpha \wedge \beta||_{L^2_{r,2\delta}} \leq C \cdot ||\alpha||_{L^2_{k,\delta}} \cdot ||\beta||_{L^2_{k',\delta}}$$

provided that the following Sobolev inequality holds:

$$r - \frac{4}{2} < \left(k - \frac{4}{2}\right) + \left(k' - \frac{4}{2}\right)$$

7.2. Spaces of connections asymptotic to flow lines for a truncation of Ξ_Γ.

In this section will define the space of connections asymptotic to a flow line for a suitable truncation of Ξ_Γ, the appropriate gauge group, and the quotient space in the based and unbased context.

First we need to fix the following data once and for all:

A1. a base point $(0, n) \in \{0\} \times N$,

A2. a constant μ with $\mu \geq \{max(\mu^+_\Gamma, \mu^-_\Gamma)\}$ for all flat connections Γ on η,

A3. a metric $g \in B$ satisfying $||g(t) - g_0(t)||_{C^1} < e^{-\mu t}$.

Next fix a C^∞ flat connection Γ on η. As in Section V.1 let $\mathcal{N}_\Gamma \subset H^1(N; ad\Gamma)$ be a $Stab(\Gamma)$-invariant neighborhood of $\{0\}$ and let $\zeta\colon \mathcal{N}_\Gamma \to H^\perp \subset \mathcal{K}_\Gamma$ be a $Stab(\Gamma)$-equivariant map whose graph is a $Stab(\Gamma)$-invariant center manifold $\mathcal{H}_\Gamma$ for ∇f_Γ. Let $Y_\Gamma \subset \mathcal{H}_\Gamma$ be the stable subset for ∇f_Γ. With respect to this choice we fix the following local data:

B1. A constant $\delta > 0$ with $\mu_\Gamma^-/2 < \delta/2 < \mu_\Gamma^-$.

B2. If necessary restrict further $\mathcal{N}_\Gamma$ so that for any pair of connections A_1, A_2 on η corresponding to points in $\mathcal{H}_\Gamma$ we have that

$$|CS(A_2) - CS(A_1)| < \epsilon_0/2$$

B3. Simon data $\mathcal{S}_\Gamma = (V_\Gamma, T_0)$

B4. $V_\Gamma \subset U_\Gamma$ a Simon patch pair such that:

(a) for every flat connection $\Gamma' \in U_\Gamma$ the constant $\delta/2$ is not in the spectrum of $- * d_{\Gamma'}|Im(d^*_{\Gamma'}(\Omega^2(N))$, and

(b) U_Γ satisfies Lemma 2.3.2.

(c) $U_\Gamma \subset \{\Gamma\} + \mathcal{N}_\Gamma \times H^\perp$.

For convenience we call a collection of such choices *local data around* Γ.

7.2.1. The truncated vector field Ξ_Γ^{tr}. Let $\beta\colon \mathcal{H}_\Gamma \to [0,1]$ be a $Stab(\Gamma)$ invariant C^2 cutoff function whose support is contained in V_Γ. Let $\mathcal{H}_{in}$ denote the interior of the set where β is identically 1. Let $\mathcal{H}_{out} = \beta^{-1}((0,1])$. Let Ξ_Γ denote the restriction of the vector field $\nabla_A f_\Gamma$ to $\mathcal{H}_\Gamma$ and we define the vector field Ξ_Γ^{tr} on $\mathcal{H}_\Gamma$ by the equation

$$\Xi_\Gamma^{tr}(h) = \beta(h)\Xi_\Gamma(h) \ for\ all\ h \in \mathcal{H}_\Gamma. \tag{79}$$

DEFINITION 7.2.1. *A cutoff function β with the above properties is called a trimming of the center manifold $\mathcal{H}_\Gamma$ compatible with V_Γ. The vector field Ξ^{tr} is called the trimmed vector field. A trimmed center manifold is one with a specified trimming.*

Note that by construction the critical set of Ξ_Γ^{tr} is equal to $Crit(\Xi_\Gamma) \cup (\mathcal{H}_\Gamma \setminus \mathcal{H}_{out})$. The vector field Ξ_Γ^{tr} has the following useful properties.

LEMMA 7.2.2.
- *The flow of Ξ_Γ^{tr} is complete.*

- *If $h, h' \in \mathcal{H}_\Gamma$ lie on the same flow line for Ξ_Γ^{tr}, then they lie on the same flow line for Ξ_Γ.*

Proof. The first item follows immediately from the fact that Ξ_Γ^{tr} is a C^2 compactly supported vector field. The second item follows from the fact that Ξ_Γ and Ξ_Γ^{tr} point in the same direction whenever $\Xi_\Gamma^{tr} \neq 0$. □

For $h \in \mathcal{H}_\Gamma$ let $h(t)$ denote the flow line for Ξ_Γ^{tr} with initial condition $h = h(T_0)$. In order to thicken the moduli space we will consider a space of connections which are exponentially asymptotic to flow lines of Ξ_Γ^{tr}. In order for this to be useful we must show that flow lines of Ξ_Γ^{tr} cannot approach each other too rapidly. Since the cutoff function $\beta \leq 1$, a flow line for Ξ_Γ^{tr} approaches any limit point more slowly than the Ξ_Γ flow line with same initial condition. More precisely we have

PROPOSITION 7.2.3. *Let $h(t)$ and $h'(t)$ be flow lines for Ξ_Γ^{tr} defined on $[T_0, \infty)$. If there is a constant $C > 0$ so that*

$$Dist(h(t), h'(t)) \leq Ce^{-\frac{\mu_\Gamma^- t}{2}}$$

for all t, then $h = h'$.

To prove this we need the following lemma.

LEMMA 7.2.4. *Fix $h \in \mathcal{H}_\Gamma$, let $h(t)$, $T_0 \leq t < \infty$ be the flow line for Ξ_Γ^{tr} with initial condition $h(T_0) = h$. Then the limit $\lim_{t\to\infty} h(t)$ exists. Furthermore, for any pair $h, h' \in \mathcal{H}_\Gamma$ if*

$$h_\infty = \lim_{t\to\infty} h(t) = \lim_{t\to\infty} h'(t)$$

then either h and h' lie on the same flow line or $h, h' \in Y_\Gamma$. (In this last case the Ξ_Γ flow lines with initial conditions h and h' exist for all $t \geq T_0$ and have the same limit as $t \mapsto \infty$.)

Proof. First we fix some notation. For $h \in \mathcal{H}_{out}$ let $a\colon [T_0, T_1) \to \mathcal{H}_{out}$ be the maximal flow line for Ξ_Γ with $a(T_0) = h$ and $a(t) \in \mathcal{H}_{out}$. Of course

$$a([T_0, T_1)) = h([0, \infty)). \tag{80}$$

Given this, the proof of the lemma is an easy exercise from the following claims.

CLAIM 7.2.5. *If $T_1 = \infty$ then the path $a(t)$ (and hence the path $h(t)$) has a unique limit and $h \in Y_\Gamma$.*

Proof. If $T_1 = \infty$ then $\omega = \Gamma + a(t) + \Theta(a(t))dt$ is a g_0-ASD connection on $[0, \infty) \times \eta$. Also we have

$$\int_{T_0}^{\infty} |F_\omega|^2 = \lim_{T \to \infty} \int_{T_0}^{T} |F_\omega|^2 = \lim_{T \to \infty} |CS(\omega|_{\{T\} \times \eta}) - CS(\omega|_{\{T_0\} \times \eta})| \le \epsilon_0/2 < \epsilon_0.$$

Since $h \in \mathcal{H}_\Gamma \subset V_\Gamma$ Theorem 4.3.1 applies to give a unique limiting flat connection Γ'. Thus the path $a(t)$ has a unique limit point. By Equation 80 the same holds for $h(t)$. □

CLAIM 7.2.6. *If $T_1 \le \infty$ then the path $a(t)$ (and hence the path $h(t)$) has a unique limit h_∞ and then $\Xi_\Gamma(h_\infty) \neq 0$.*

Proof. If $T_1 \neq \infty$ then since Ξ is a bounded vector field on $\mathcal{H}_\Gamma$ the limit $\lim_{t \to T_1} a(t) = a_\infty$ exists and is unique. Again Equation 80 implies the corresponding result for $h(t)$. □

□

Now let us prove Proposition 7.2.3. Clearly if the flow lines corresponding to h and h' are exponentially asymptotic then they have the same limit. Thus either both flow lines converge to Y_Γ, in which case the result follows from Lemma 5.3.3, or h and h' lie on the same flow line and converge to h_∞ where $\Xi(h_\infty) \neq 0$ and $\beta(h_\infty) = 0$. This case is elementary.

7.2.2. The space of connections. Fix a set of local data around Γ and a trimming $(\mathcal{H}_\Gamma, \beta)$ for its center manifold. We call all of this *thickening data centered at* Γ and denote it by $\mathcal{T}_\Gamma$. For each $h \in \mathcal{H}_{out}$ there is a unique flow line $h(t)$, $T_0 \leq t < \infty$, in U_Γ for the vector field Ξ_Γ^{tr} with $h(T_0) = h$. For any $h \in \mathcal{H}_{out}$ let $\hat{\omega}_h = h(t) + \Theta(h(t))dt$. Note that while $\hat{\omega}_h$ is not in general g_0-ASD, it is g_0-ASD in neighborhood of any slice $\{t\} \times \eta$ where $h(t) \in \mathcal{H}_{in}$.

Let $\mathcal{A}_w(\mathcal{T}_\Gamma, T_0)$ denote the space of connections ω on E so that the following hold:

(1) $\omega \in L^2_{2,loc}$,
(2) There is an $h \in \mathcal{H}_{out}$ so that $\omega - \hat{\omega}_h \in L^2_{2,\delta}([T_0, \infty))$, and
(3) for each $t \geq T_0$ the connection $\omega|(\{t\} \times \eta)$ is gauge equivalent to a connection in U_Γ.

By Proposition 7.2.3, for all distinct pairs $h, h' \in \mathcal{H}_{out}$ we have that $\hat{\omega}_h - \hat{\omega}_{h'}$ is never in $L^2_{2,\delta}([T_0, \infty) \times \eta)$ hence any $\omega \in \mathcal{A}_w(\mathcal{T}_\Gamma, T_0)$ determines a unique $\hat{\omega}_h$ and hence $h \in \mathcal{H}_{out}$ and hence a map $\tilde{p} \colon \mathcal{A}_w(\mathcal{T}_\Gamma, T_0) \to \mathcal{H}_{out}$. For any $t \geq T_0$ we define a map

$$\tilde{p}_t : \mathcal{A}_w(\mathcal{T}_\Gamma, T_0) \to \mathcal{H}_{out}$$

by setting $\tilde{p}_t(\omega)$ equal to the value at t of the flow line γ for Ξ^{tr} with initial conditions $\gamma(T_0) = \tilde{p}_{T_0}(\omega)$. We denote $\tilde{p}_{T_0}$ by $\tilde{p}$.

We topologize $\mathcal{A}_w(\mathcal{T}_\Gamma, T_0)$ by defining a neighborhood basis indexed by the following data. Fix a compact set $K \subset X$ so that $K \cup ([T_0, \infty) \times N) = X$, a neighborhood $\mathcal{O}$ of $\omega|_K$ in $L^2_2(K)$, a neighborhood $\mathcal{O}'$ of $\omega - \hat{\omega}_h$ in $L^2_{2,\delta}([T_0, \infty) \times N)$ and a neighborhood $\mathcal{O}''$ of $h \in \mathcal{H}_{out}$. The corresponding neighborhood of ω in $\mathcal{A}_w(\mathcal{T}_\Gamma, T_0)$ is the set of all ω' so that $\omega'|_K \in \mathcal{O}$, $\omega' - \hat{\omega}_{h',T_0} \in \mathcal{O}'$ and $h' \in \mathcal{O}''$. With this topology, $\mathcal{A}_w(\mathcal{T}_\Gamma, T_0)$ is a C^2 Hilbert manifold and the projection p is a smooth map. Note that $\mathcal{A}_w(\mathcal{T}_\Gamma, T_0) \subset \mathcal{A}_w(\mathcal{T}_\Gamma, T_0')$ whenever $T_0 \leq T_0'$.

We form the union $\mathcal{A}_w(\mathcal{T}_\Gamma) = \cup_{T_0 \geq 1} \mathcal{A}_w(\mathcal{T}_\Gamma, T_0)$ and give this space the direct limit topology. It is also a C^2-Hilbert manifold.

Since all flat connections in $\mathcal{H}_\Gamma$, are in the same component of the character variety as Γ, we can divide $\mathcal{A}_w(\mathcal{T}_\Gamma)$ into components $\mathcal{A}_{\ell,w}(\mathcal{T}_\Gamma)$ indexed by the energy ℓ so that for all $\omega \in \mathcal{A}_{\ell,w}(\mathcal{T})$ we have $|c_2(\omega) - \ell| < \epsilon_0/2$. The possible values for ℓ differ from each other by integers.

7.2.3. The gauge group and the quotient space. Let $\mathcal{G}(\mathcal{T}_\Gamma, T_0)$ be the the group of bundle automorphisms of E with the property that for all $\omega \in \mathcal{A}_w(\mathcal{T}_\Gamma, T_0)$ we have $\sigma^*\omega \in \mathcal{A}_w(\mathcal{T}_\Gamma, T_0)$. We say that σ is an $L^2_{3,\delta}$-gauge transformation if σ agrees with the exponential of an $L^2_{3,\delta}$-section of adE on the end of X.

LEMMA 7.2.7. *$\mathcal{G}(\mathcal{T}_\Gamma, T_0)$ is equal to the set of $L^2_{3,loc}$ gauge transformations σ so that there is τ_0, the pull back of an element of $Stab(\Gamma)$, with $\sigma|_{[0,\infty)\times\eta} \circ \tau_0$ an $L^2_{3,\delta}$ gauge transformation.*

Proof. We assume for simplicity that G is a linear group, i.e that G has a faithful representation into $SU(n)$ for some n. Form the associated bundle $\xi = E \times_G M_n(C)$ where G acts by conjugation. We can define the gauge group $\mathcal{G}_\delta$ to be the space of sections σ of ξ so that for all $b \in X$, $\sigma(b) \in G$ and so that $\sigma - Id \in L^2_{3,\delta}$.

First we prove that any element σ of the stabilizer of $\omega \in \mathcal{A}(\mathcal{T}_\Gamma)$ is of the required form. On $[T_0, \infty) \times \eta$ write $\omega = \hat{\omega}_h + a$. Then $\sigma^*\omega = \omega$ becomes

$$d_{\hat{\omega}_h}(\sigma^{-1}) = \sigma^{-1}a - a\sigma^{-1}$$

where we view σ as a section of ξ. Note that the righthand side of this equation is in L^2_δ. It follows directly from Lemma 5.2 of [26] that σ is the product of a $\hat{\omega}_h$ covariantly constant gauge transformation and an $L^2_{3,\delta}$ gauge transformation.

The fact that $\mathcal{G}(\mathcal{T}_\Gamma, T_0)$ is a subgroup of the group of $L^2_{3,loc}$ gauge transformation is standard (cf. [12].) Now choose $\omega, \omega' \in \mathcal{A}_w(\mathcal{T}_\Gamma, T_0)$ and $\sigma \in \mathcal{G}(\mathcal{T}_\Gamma, T_0)$ so that $\sigma^*\omega = \omega'$. Let h, h' denote the corresponding points in $\mathcal{H}_{out}$. According to Lemma 3.1.3 for some T_0 sufficiently large there are gauge transformations $\sigma_1, \sigma_1' \in \mathcal{G}_{(1,2)}([T_0, \infty) \times \eta)$ putting $\omega|_{[T_0,\infty)\times\eta}$ and $\omega'|_{[T_0,\infty)\times\eta}$ in to standard form with respect to V_Γ. Furthermore by Corollary 3.1.7 these gauge transformations can be chosen to lie in $\mathcal{G}_{(1,2),\delta}([T_0,\infty)\times\eta)$. Since $\sigma_1^*(\omega)$ and $\sigma_1'^*(\omega')$ are gauge equivalent connections in standard form they are gauge equivalent by a τ_0 which is the pullback of an element in $Stab(\Gamma)$. Thus there is an element σ'' of $Stab(\omega|_{[T_0,\infty)\times\eta})$ $\sigma = \sigma_1'^{-1}\tau_0^{-1}\sigma_1'\sigma''$. The result follows directly from this last equation and the first paragraph of the proof. □

COROLLARY 7.2.8. *$\mathcal{G}(\mathcal{T}_\Gamma, T_0)$ depends only on Γ and δ.*

Because of this corollary we set $\mathcal{G}_\delta(\Gamma) = \mathcal{G}(\mathcal{T}_\Gamma, T_0)$.

We note that the proof of the above lemma has given us a little bit more.

COROLLARY 7.2.9. *For all pairs of gauge equivalent connections $\omega, \omega' \in \mathcal{A}(\mathcal{T}_\Gamma, T_0)$ with $p(\omega) = h$ and $p(\omega') = h'$ there is a $\tau \in Stab(\Gamma)$ so that $\tau \cdot h = h'$.*

We can now form the quotient space.

DEFINITION 7.2.10. $\mathcal{B}_{\ell,w}(\mathcal{T}_\Gamma, T_0) = \mathcal{A}_{\ell,w}(\mathcal{T}_\Gamma, T_0)/\mathcal{G}_\delta(\Gamma)$

There is also a based version of the quotient space.

DEFINITION 7.2.11. $\mathcal{B}^0_{\ell,w}(\mathcal{T}_\Gamma, T_0) = \mathcal{A}_{\ell,w}(\mathcal{T}_\Gamma, T_0) \times_{\mathcal{G}_\delta(\Gamma)} \eta_n$

Corollary 7.2.9 implies that the projection $p\colon \mathcal{A}_{\ell,w}(\mathcal{T}_\Gamma, T_0) \to \mathcal{H}_\Gamma$ descends to define a continuous map $p\colon \mathcal{B}_{\ell,w}(\mathcal{T}_\Gamma, T_0) \to \mathcal{H}_\Gamma/Stab(\Gamma)$ and a continuous $SO(3)$ equivariant map $p^0 : \mathcal{B}^0_{\ell,w}(\mathcal{T}_\Gamma, T_0) \to \mathcal{H}_\Gamma \times_{Stab(\Gamma)} \eta_n$ We will eventually prove a slice theorem for the gauge group action. This will imply that $\mathcal{B}^0_{\ell,w}(\mathcal{T}_\Gamma, T_0)$ is a C^2 Hilbert manifold and that p^0 is a C^2 map.

7.2.4. Connections δ-decaying to flat connections. For each $\delta > 0$ there is a globally defined space which for appropriate δ is identified with the preimage of the flat connections in $\mathcal{B}^0_{\ell,w}(\mathcal{T}_\Gamma, T_0)$. It is the purpose of this subsection to define this space.

Fix $\delta > 0$. Consider the subspace $\mathcal{B}^0_{\ell,w,\delta}(X)$ of $L^2_{3,loc}$ gauge equivalence classes of based $L^2_{2,loc}$ connections on E with representatives which, with respect to some product structure on the end, differ from the pullback of a flat connection by a 1-form in $L^2_{2,\delta}$ and with energy $8\pi^2\ell$. We shall give this space a finer topology than the $L^2_{2,loc}$ topology, one that is analogous to the topologies above.

A neighborhood base for the topology on $\mathcal{B}^0_{\ell,w,\delta}(X)$ is given as follows. Let (ω, f) represent a point of $\mathcal{B}^0_{\ell,w,\delta}(X)$. Let $\phi\colon [0,\infty) \times \eta \to E|([0,\infty) \times N)$ be a product structure and let Γ be a flat connection on η so that $\phi^*(\omega) - \pi^*(\Gamma) \in L^2_{2,\delta}$. A basis of neighborhoods is given by choosing open neighborhoods: U_1 of $\omega|X_1$ in the L^2_2 topology; U_2 of $\phi^*(\omega) - \pi^*(\Gamma)$ in the $L^2_{2,\delta}$ topology; and U_3 of $[\Gamma, f]$ in $\mathcal{R}(N)$. The neighborhood in $\mathcal{B}^0_{\ell,w,\delta}(X)$ determined by these choices of the class of $[\omega, f]$ is the set of gauge equivalence classes $[\omega', f']$ such that $\omega'|X_1 \in U_1$, and so that there is a flat connection Γ' with $[\Gamma', f'] \in U_3$ and $\phi^*(\omega') - \pi^*(\Gamma') \in U_2$.

There is a continuous map $p\colon \mathcal{B}^0_{\ell,w,\delta}(X) \to \mathcal{R}(N)$. Fix thickening data $\mathcal{T}_\Gamma$ and let δ be the constant of the thickening data. Let $C \subset \mathcal{H}_{in}$ denote the critical set of ∇f_Γ. Then $C \times_{Stab\Gamma} \eta_n$ is identified with an open subset of $\mathcal{R}(N)$. Under this identification for $T_0 \geq 1$ the preimage under p_{T_0} of $C \times_{Stab(\Gamma)} \eta_n$ is identified with a subset of the preimage under p of $C \times_{Stab(\Gamma)} \eta_n$.

The slice theorem mentioned in the above section will be used to show that $\mathcal{B}^0_{\ell,w,\delta}(X)$ is a stratified space with local cone bundle neighborhoods fibering over the representation variety.

7.2.5. The map to the space of $L^2_{2,loc}$ connections modulo gauge equivalence. On the level of connections there is an obvious continuous injection $\iota\colon \mathcal{A}_w(\mathcal{T}_\Gamma) \to \mathcal{A}_{2,loc}(E)$. It is easy to see that this map is not a homeomorphism onto its image. Lemma 7.2.7 implies that this map descends to define a continuous injection $\iota\colon \mathcal{B}_{\ell,w}(\mathcal{T}_\Gamma) \to \mathcal{B}_{2,loc}(E)$. Similarly on the based level there is a continuous injection $\iota\colon \mathcal{B}^0_{\ell,w}(\mathcal{T}_\Gamma) \to \mathcal{B}^0_{2,loc}(E)$.

7.3. The thickened moduli space associated to a truncation of Ξ.

Let us fix thickening data $\mathcal{T}_\Gamma$ and $T_0 \geq 1$. We consider the following modification of the Anti-Self-Dual Yang-Mills equation. Let $s\colon \mathcal{A}_w(\mathcal{T}_\Gamma, T_0) \to \Omega^2_{+,1,loc}$ be defined by the following formula:

$$s(\omega) = P_+(g)(F(\omega) - \beta' P_+(g_0)F(\hat{\omega}_h)).$$

where $p(\omega) = h \in \mathcal{H}_{in}$ and where $\beta'\colon X \to [0,1]$ is smooth cutoff function with $\beta'(x) = 0$ if $\tau(x) \leq T_0$ and $\beta'(x) = 1$ if $\tau(x) \geq T_0 + 1$. The equation $s(\omega) = 0$ is called the *modified Anti-Self-Dual Yang-Mills Equation* or mASD for short.

LEMMA 7.3.1. *The image of s is contained in $\Omega^2_{+,1,\delta}$ and the resulting map*

$$s\colon \mathcal{A}_w(\mathcal{T}_\Gamma, T_0) \to \Omega^2_{+,1,\delta}$$

is a C^1 map.

Proof. Write $\omega|_{[T_0,\infty)\times\eta} = \hat{\omega}_h + \alpha$. Expanding the equation for s we have on $[T_0,\infty) \times \eta$.

$$\begin{aligned} s(\omega) &= P_+(g)(F_{\hat{\omega}_h} + d_{\hat{\omega}_h}\alpha + \frac{1}{2}\alpha\wedge\alpha) - \beta' P_+(g_0)F_{\hat{\omega}}) \\ &= P_+(g)(d_{\hat{\omega}_h}\alpha + \frac{1}{2}\alpha\wedge\alpha)) + \\ &\quad P_+(g)(P_+(g) - \beta' P_+(g_0))(F_{\hat{\omega}_h}). \end{aligned}$$

Since $\hat{\omega}_h$ is bounded in uniform C^2 topology, $|g(t,n) - g_0(t,n)|_{C^1} \leq Ce^{-\mu t}$, and since $\alpha \in L^2_{2,\delta}$ it is clear from this last expression that the image of s is in $L^2_{1,\delta}$. The smoothness of s follows from the Sobolev multiplication theorems and the fact that the map sending $h \mapsto \hat{\omega}_h$ is C^2. □

We now define the thickened moduli space. Let

$$\mathcal{M}_{\ell,w}(\mathcal{T}_\Gamma, T_0, g) = \{\omega \in \mathcal{A}_{\ell,w}(\mathcal{T}_\Gamma, T_0) | s(\omega) = 0\}/\mathcal{G}_\delta(\Gamma).$$

For all $t \geq T_0$ there are the projection maps

$$p_t \colon \mathcal{M}_{\ell,w}(\mathcal{T}_\Gamma, T_0, g) \to \mathcal{H}_\Gamma / Stab(\Gamma)$$

induced by the $\tilde{p}_t$. We also denote p_{T_0} by p.

We also have the based thickened moduli space

$$\mathcal{M}^0_{\ell,w}(\mathcal{T}_\Gamma, T_0, g) = \{(\omega, f) \in \mathcal{A}_{\ell,w}(\mathcal{T}_\Gamma, T_0, g) \times \eta_n | s(\omega) = 0\}/\mathcal{G}_\delta(\Gamma)$$

and for every $t \geq T_0$ the corresponding projections

$$p^0_t \colon \mathcal{M}^0_{\ell,w}(\mathcal{T}_\Gamma, T_0, g) \to \mathcal{H}_\Gamma \times_{Stab(\Gamma)} \eta_n$$

defined by

$$p^0_t([\omega, e]) = [\tilde{p}_t(\omega), f_t]$$

where f_t is the frame in η_n obtained form e by parallel transport using ω along $[0, t] \times \{n\}$. We set $p^0 = p^0_{T_0}$.

7.4. The relationship between the L^2-moduli space and the thickened moduli space

Fix a Simon patch pair $V'_\Gamma \subset U'_\Gamma$ centered at Γ so that $U'_\Gamma \subset \{\Gamma\} + \mathcal{H}_{in} \times H^\perp$. Let $\mathcal{S}'_\Gamma = (V'_\Gamma, T_0)$ be Simon data.

DEFINITION 7.4.1. *Let $\mathcal{M}_{\ell,w,\delta}(Y_\Gamma, \mathcal{S}'_\Gamma, g) \subset \mathcal{M}_{\ell,w}(\mathcal{T}_\Gamma, T_0, g)$ denote the subspace of mASD connections so that $p([\omega]) \in Y_\Gamma \cap V'_\Gamma$, Equation 49 holds for $T_1 = \infty$ and $\omega | \{T_0\} \times \eta$ is $L^2_{\frac{5}{2}}$ gauge equivalent into V'_Γ. This is our local model of the L^2-moduli space. There is also the based version $\mathcal{M}^0_{\ell,w,\delta}(Y_\Gamma, \mathcal{S}'_\Gamma, g)$.*

LEMMA 7.4.2. *Every g-mASD connection in $\mathcal{M}_{\ell,w,\delta}(Y_\Gamma, \mathcal{S}'_\Gamma, g)$ is in fact g-ASD. Furthermore $\mathcal{M}_{\ell,w,\delta}(Y_\Gamma, \mathcal{S}'_\Gamma, g)$ is the preimage of $\mathcal{M}_{\ell,w}(\mathcal{S}'_\Gamma, g)$ under the injection $\mathcal{B}_{\ell,w}(\mathcal{T}_\Gamma, T_0) \to \mathcal{B}_{2,loc}(E)$ with the same statement holding in the based context.*

Proof. By Theorem 4.3.1 for all $t \geq T_0$ the flow line corresponding to $p(\omega)$ remains inside U'_Γ and hence inside $\mathcal{H}_{in}$. Thus the second term in the definition of s is zero. The second statement is obvious. □

It remains to show that our model space describes an open set in the moduli space.

THEOREM 7.4.3. *The injection* $\iota\colon \mathcal{B}^0_{\ell,w}(\mathcal{T}_\Gamma, T_0) \to \mathcal{B}^0_{2,loc}(E)$ *induces a G-equivariant homeomorphism*

$$i\colon \mathcal{M}^0_{\ell,w,\delta}(Y_\Gamma, \mathcal{S}'_\Gamma, g) \to \mathcal{M}^0_{\ell,w}(\mathcal{S}'_\Gamma, g).$$

Furthermore, the following diagram commutes.

$$\begin{array}{ccccc} \mathcal{M}^0_{\ell,w}(\mathcal{T}_\Gamma, T_0, g) & \xleftarrow{\supset} & \mathcal{M}^0_{\ell,w,\delta}(Y_\Gamma, \mathcal{S}'_\Gamma, g) & \xrightarrow{i} & \mathcal{M}^0_{\ell,w}(\mathcal{S}'_\Gamma, g) \\ & & {\scriptstyle p}\Big\downarrow & & \Big\downarrow{\scriptstyle \Upsilon_{S'}} \\ \mathcal{H}_\Gamma \times_{Stab(\Gamma)} \eta_n & \xleftarrow{\supset} & Y_\Gamma \times_{Stab(\Gamma)} \eta_n & \xrightarrow{=} & Y_\Gamma \times_{Stab(\Gamma)} \eta_n. \end{array}$$

Proof. By Lemma 7.4.2 every connection representing an element in $\mathcal{M}^0_{\ell,w,\delta}(Y_\Gamma, \mathcal{S}'_\Gamma, g)$ is ASD. Thus, $Im(i) \subset \mathcal{M}^0_{\ell,w}(X, g)$, and hence is contained in $\mathcal{M}^0_{\ell,w}(\mathcal{S}', g)$. Since ι is continuous, injective and G-equivariant so is i. It remains to verify that i is surjective and is an open mapping. For this we construct the inverse map

$$j\colon \mathcal{M}^0_{\ell,w}(\mathcal{S}'_\Gamma, g) \to \mathcal{M}^0_{\ell,w,\delta}(Y_\Gamma, \mathcal{S}'_\Gamma, g).$$

By Theorem 4.3.1 we can represent any point in $\mathcal{M}^0_{\ell,w}(\mathcal{S}'_\Gamma, g)$ by a pair (ω, f) where ω is a C^3 g-ASD connection with $\omega|(T_0,\infty)\times\eta) \in Std(\mathcal{S}'_\Gamma, g)$ and where f is a frame in $E|(0,n)$. Let $h = \Upsilon_{\mathcal{S}'_\Gamma}(\omega|_{[T_0,\infty)\times\eta})$. We write $\omega|_{[T_0,\infty)\times\eta} = \hat{\omega}_h + \alpha$. It follows immediately from Theorem 5.2.2 that $\omega \in \mathcal{A}_{\ell,w}(\mathcal{T}_\Gamma, T_0)$ and that $\alpha \in \Omega^1_{2,\delta}(X)$ and depends continuously on $\omega|_{[T_0,\infty)\times\eta} \in Std(\mathcal{S}'_\Gamma, g)$.

We define $j([\omega, f]_{loc}) = [\omega, f]_\delta$ where $[\cdot]_{loc}$ denotes the $\mathcal{G}_{3,loc}$ gauge equivalence class while $[\cdot]_\delta$ denotes the $\mathcal{G}_\delta(\Gamma)$ gauge equivalence class. The map j is well defined since according to Corollary 3.1.4 any gauge equivalence between connections in standard form on the end of X is the pullback of an element of $Stab(\Gamma)$ on the end of X.

The continuity of j follows from the assertion in Theorem 5.2.2 regarding the continuity of $\Upsilon_{\mathcal{S}'_\Gamma}$ and the $L^2_{2,\delta}$ continuity of the α.

It is clear from the definition that $i \circ j$ is the identity on $\mathcal{M}^0_{\ell,w}(\mathcal{S}'_\Gamma, g)$. Since i is injective, it then follows that $j \circ i$ is also the identity. Hence, j is an inverse homeomorphism to i.

It is clear that j commutes with the maps to Y_Γ. Consequently, so does i. □

7.5. The μ-map for the thickened moduli space

Fix thickening data $\mathcal{T}_\Gamma, T_0 > 0, g \in B$ and $\ell > 0$. Let $\mathcal{A}^*$ denote the subspace of $\mathcal{A}_{\ell,w}(\mathcal{T}_\Gamma, T_0)$ consisting of irreducible connections and let $\mathcal{B}^*$ denote the subspace of $\mathcal{B}_{\ell,w}(\mathcal{T}_\Gamma, T_0)$ consisting of irreducible orbits. Of course if Γ is irreducible then there are no reducibles in $\mathcal{A}_{\ell,w}(\mathcal{T}_\Gamma, T_0)$. Consider the quotient space $\mathcal{P} = \mathcal{A}^* \times E/\mathcal{G}_\delta(\Gamma)$. The projection $p\colon \mathcal{P} \to \mathcal{B}^* \times X$ is a principal $G/center(G)$-bundle. We define as in the closed case $\mu\colon H_2(X; Z) \to H^2(\mathcal{B}^*; Z)$ by

$$\mu(\alpha) = -\frac{1}{4}p_1(\mathcal{P})/\alpha.$$

The next task is to show that just as in the closed case there are nice geometric representatives for $\mu(\alpha)$ on the thickened moduli space $\mathcal{M}_{\ell,w}(\mathcal{T}_\Gamma, T_0, g)$.

The following lemma is the analogue for manifolds with cylindrical ends of the discussion on page 355-356 of [10].

LEMMA 7.5.1. *Fix $\Sigma \subset X_{T_0}$ a closed embedded oriented riemann surface with $[\Sigma] = \alpha$ so that $H_1(\Sigma; Z) \to H_1(X; Z)$ is a surjection. For a generic perturbation $\iota\colon \Sigma \to X$ of Σ the restriction map $\iota^*\colon \mathcal{M}_{\ell,w}(\mathcal{T}_\Gamma, T_0, g) \to \mathcal{B}(\Sigma')$ carries irreducible connections to irreducible connections.*

Proof. First we establish the following.

CLAIM 7.5.2. *Let ω be an mASD connection on E. If ω is reducible on a ball $B \subset X_{T_0}$ and on Σ then ω reduces to an $U(1)$ connection.*

Proof. The unique continuation result for the ASD equation implies directly that ω is reducible to an $O(2)$ connection on X_{T_0}. After putting ω into standard form $A + \alpha dt$ on $[T_0, \infty) \times \eta$ we see that A satisfies a flow equation of Γ and has the unique continuation property on the tube as well. Thus ω is reducible to an $O(2)$ connection on all of X. Since ω is reducible to $U(1)$ on Σ and $H_1(\Sigma; Z)$ maps onto $H_1(X; Z)$, we must have that ω reduces to $U(1)$ on X. □

Let us return to the proof of the lemma. Let $\mathcal{E}$ be a neighborhood of $\Sigma \hookrightarrow X$ in the space of C^1 embeddings ι of Σ into X. For any irreducible connection ω let us show that the subset $\mathcal{E}' \subset \mathcal{E}$ of embeddings so that $\iota^*(\omega)$ is irreducible is open and dense. Openness is clear. Let ι be an embedding so that $\iota^*(\omega)$ is reducible. Let B be a ball in X intersecting Σ. Since ω is irreducible on the ball we can find arbitrarily small C^1 perturbations of ι supported in the ball so that $\iota^*(\omega)$ is irreducible. Since the topology of the moduli space has a countable base the desired result follows from this and the Baire category theorem. □

Fix a compact riemann surface Σ and a spin structure on Σ. Let $\mathcal{B}^*(\Sigma)$ denote the space of gauge equivalence classes of $SU(2)$ connections on a principal $SU(2)$ bundle over Σ and let $\mathcal{L}_\Sigma \to \mathcal{B}^*(\Sigma)$ denote the determinant line bundle of the Dirac operator coupled to connections. The first Chern class of this line bundle is $\mu([\Sigma]) \in H^2(\mathcal{B}^*(\Sigma))$.

LEMMA 7.5.3. *Given any homology class $\alpha \in H_2(X; Z)$ there is an embedded riemann surface $\Sigma \subset X$ representing α such for a generic perturbation $\iota \colon \Sigma \to X$ of Σ, we have that ι^* maps the thickened moduli space to $\mathcal{B}^*(\Sigma)$ and the first Chern class of the line bundle $\iota^*\mathcal{L}_\Sigma$ is the restriction of $\mu(\alpha)$ to the thickened moduli space.*

Proof. This is immediate from the previous lemma and the naturality of the μ-map. $\square$

CHAPTER 8

Structure of the fibers of $p^0\colon \mathcal{M}^0_{\ell,w}(\mathcal{T}_\Gamma, T_0, g) \to \mathcal{H}^0_\Gamma$

In the previous chapter we introduced the based thickened moduli space $\mathcal{M}^0_{\ell,w}(\mathcal{T}_\Gamma, T_0, g)$ and the map p^0 of this space to the based center manifold $\mathcal{H}^0_\Gamma = \mathcal{H}_\Gamma \times_{Stab(\Gamma)} \eta_n$. This chapter has two related goals. First, we study the fibers of the map p^0 in more detail. Here we have a situation that is essentially identical to the δ-decaying moduli space picture. The structure of the fibers is thus analyzed as in [26]. Second, we study the structure of the thickened moduli space as a finite dimensional modification of the central fiber. This gives us the local structure of the based thickened moduli space and hence of based L^2-moduli space. We adopt all the notation and choices of the previous chapter.

8.1. The fiber over a point

Fix $h \in \mathcal{H}_{out}$. We define the space $\mathcal{A}_{\ell,w}(h, \mathcal{T}_\Gamma, T_0)$ as the fiber of $p\colon \mathcal{A}_{\ell,w}(\mathcal{T}_\Gamma, T_0) \to \mathcal{H}_\Gamma$ over h. Clearly, $\mathcal{A}_{\ell,w}(h, \mathcal{T}_\Gamma, T_0)$ is an affine space for $\Omega^1_{2,\delta}(X)$.

DEFINITION 8.1.1. *We define the based gauge group $\mathcal{G}_\delta$ the group of all $L^2_{3,\delta}$-gauge transformations. This is a normal subgroup of $\mathcal{G}_\delta(\Gamma)$ with quotient $Stab(\Gamma)$. Its Lie algebra is $\Omega^0_{3,\delta}(X)$.*

The differential of the action of the based gauge group restricted to

$\mathcal{A}_{\ell,w}(h,\mathcal{T}_\Gamma,T_0)$ at ω is

$$d_\omega: \Omega^0_{3,\delta}(X) \to \Omega^1_{2,\delta}(X).$$

We define $\mathcal{B}_{\ell,w}(h,\mathcal{T}_\Gamma,T_0)$ to be the quotient of this action. There is an action of $Stab(h)$ on $\mathcal{B}_{\ell,w}(h,\mathcal{T}_\Gamma,T_0)$ with quotient identified with the fiber of $p\colon \mathcal{B}_{\ell,w}(\mathcal{T}_\Gamma,T_0) \to \mathcal{H}_\Gamma/Stab(\Gamma)$ over $[h]$, and, as we will prove in Chapter 10, it is a C^2-Hilbert manifold which near $[\omega]$ is locally modeled on $Ker(e^{-\tau\delta}d^*_\omega e^{\tau\delta})$. We define $\mathcal{M}_{\ell,w}(h,\mathcal{T}_\Gamma,T_0,g) \subset \mathcal{B}_{\ell,w}(h,\mathcal{T}_\Gamma,T_0)$ to be the subset of equivalence classes of g-mASD connections. This subspace is invariant under the action of $Stab(h) \subset Stab(\Gamma)$.

As we shall see s defines a Fredholm section of the vector bundle

$$\mathcal{A}_{\ell,w}(h,\mathcal{T}_\Gamma,T_0) \times_{\mathcal{G}_\delta} \Omega^2_{+,1,\delta}(X) \to \mathcal{B}_{\ell,w}(h,\mathcal{T}_\Gamma,T_0).$$

As a result $\mathcal{M}_{\ell,w}(h,\mathcal{T}_\Gamma,T_0,g)$ has a Zariski tangent space at every point and is a smooth submanifold at $[\omega]$ if and only if s is transverse at $[\omega]$, or equivalently if and only if

$$P_+d_\omega: \Omega^1_{2,\delta}(X) \to \Omega^2_{+,1,\delta}(X)$$

is onto.

8.2. The deformation complex for the fibers of p and expansions for the δ-decaying operators

We fix a C^∞ flat connection Γ, Simon data $\mathcal{S}_\Gamma = (V_\Gamma, T_0)$, and thickening data $\mathcal{T}_\Gamma$ with all the notation as in the previous section. Write $h(t)$, $T_0 \le t < \infty$, for the flow line for Ξ^{tr}_Γ with $h(T_0) = h$. Let $h_\infty \in U_\Gamma$ be the asymptotic value of $h(t)$.

As we have seen in Lemma 7.3.1 the modified ASD equation defines a map

$$s\colon \mathcal{A}_{\ell,w}(h,\mathcal{T}_\Gamma,T_0) \to \Omega^2_{+,1,\delta}(X).$$

The differential of s at ω equal to $P_+d_\omega: \Omega^1_{2,\delta} \to \Omega^2_{+,1,\delta}(X)$. Thus associated to each g-mASD connection $\omega \in \mathcal{A}_{\ell,w}(h,\mathcal{T}_\Gamma,T_0)$ we have the deformation complex:

$$F_\delta(\omega): \qquad \Omega^0_{3,\delta}(X) \xrightarrow{d_\omega} \Omega^1_{2,\delta}(X) \xrightarrow{P_+d_\omega} \Omega^2_{+,1,\delta}(X).$$

The formal or Zariski tangent space to $\mathcal{M}_{\ell,w}(h, \mathcal{T}_\Gamma, \mathcal{T}_0, g)$ at $[\omega]$ is $H^1(F_\delta(\omega))$, the first cohomology of the complex $F_\delta(\omega)$. The condition that $\mathcal{M}_{\ell,w}(h, \mathcal{T}_\Gamma, \mathcal{T}_0, g)$ be smooth at $[\omega]$, i.e. that every Zariski tangent vector can be integrated to give the germ of a curve in $\mathcal{M}_{\ell,w}(h, \mathcal{T}_\Gamma, \mathcal{T}_0, g)$, is that $H^2(F_\delta(\omega)) = 0$. Note that $H^0(F_\delta(\omega)) = 0$ since there are no non-zero δ-decaying covariantly constant sections of adE.

In order to understand the Fredholm properties of $F_\delta(\omega)$, we find it convenient to "wrap up" the complex into a single operator by taking adjoints with respect to the L^2_δ-inner product. Thus, we define

$$D_{\omega,\delta} = (e^{-\tau\delta} d_\omega^* e^{\tau\delta}, P_+ d_\omega) \colon \Omega^1_{2,\delta}(X) \to \Omega^0_{1,\delta}(X) \oplus \Omega^2_{+,1,\delta}(X).$$

If ω is g-mASD then the kernel of this operator is $H^1(F_\delta(\omega))$ and the cokernel is identified with $H^0(F_\delta(\omega)) \oplus H^2(F_\delta(\omega)) = H^2(F_\delta(\omega))$.

On $[0,\infty) \times N$ we define the isomorphism

$$i \colon \pi^* \Big((\Lambda^0(T^*N) \oplus T^*N) \otimes adE|_N \Big) \to T^*([0,\infty) \times N) \otimes adE$$

by $i(a,b) = adt + b$, where t denotes the natural coordinate on R and $\pi \colon [0,\infty) \times N \to N$ denotes the projection. The symbol of $D_{\omega,\delta}$ evaluated on dt gives an isomorphism

$$\sigma \colon T^*([0,\infty) \times N) \otimes adE \to \\ \Big(\Lambda^0(T^*([0,\infty) \times N)) \otimes adE\Big) \oplus \Big(\Lambda^2_+(T^*([0,\infty) \times N)) \otimes adE\Big).$$

Using these isomorphisms we may consider the operator

$$i^{-1}\sigma^{-1} D_{\omega,\delta} i \colon L^2_{1,\delta}\Big([0,\infty) \times N; \pi^*((\Lambda^0(T^*N) \oplus T^*N) \otimes adE|_N)\Big) \to \\ L^2_\delta\Big([0,\infty) \times N; \pi^*((\Lambda^0(T^*N) \oplus T^*N) \otimes adE|_N)\Big).$$

This operator has the form

$$\frac{\partial}{\partial t} + A_{h_\infty,\delta} + o(1)$$

where

$$A_{h_\infty,\delta} \colon L^2(\Lambda^0 \oplus \Lambda^1(T^*N)) \to L^2(\Lambda^0 \oplus \Lambda^1(T^*N))$$

is given by the matrix of operators

$$\begin{pmatrix} \delta & -d^*_{h_\infty} \\ -d_{h_\infty} & *d_{h_\infty} \end{pmatrix}.$$

8.3. Fredholm theory of the operators $D_{\omega,\delta}$

In the last section we showed that for any $\omega \in \mathcal{A}_{\ell,w}(h, \mathcal{T}_\Gamma, T_0)$, with appropriate bundle isomorphisms over $[0,\infty) \times N$, the operator $D_{\omega,\delta}$ is given by

$$\frac{\partial}{\partial t} + A_{h_\infty,\delta} + o(1)$$

where $h_\infty \in \mathcal{H}_\Gamma$ is the limiting connection at infinity for ω. Our goal in this section is to show that the index of $D_{\omega,\delta}$ is independent of ω and to identify it with an index associated to the central flat connection Γ.

Let us start by commuting the spectrum of $A_{\Gamma,\delta}$. This spectrum can be described in terms of the Hodge decomposition $\Omega^0 \oplus \Omega^1(T^*N \otimes adE|_N) = H^0(N;\Gamma) \oplus Im(d^*_\Gamma) \oplus Im(d_\Gamma) \oplus H^1(N;\Gamma) \oplus Im(d^*_\Gamma)$ as follows. In terms of this decomposition this operator has the form

$$\begin{pmatrix} \delta & 0 & 0 & 0 & 0 \\ 0 & \delta & -d^*_\Gamma & 0 & 0 \\ 0 & -d_\Gamma & 0 & 0 & 0 \\ 0 & 0 & 0 & 0 & 0 \\ 0 & 0 & 0 & 0 & *d_\Gamma \end{pmatrix}.$$

From this the spectrum $A_{\Gamma,\delta}$ can be seen to be unbounded, but discrete without accumulation points and with each eigenvalue occurring with finite multiplicity. An eigenvalue of $A_{\Gamma,\delta}$ is of one of three types:

(type i) Each Δ_Γ-harmonic function is an eigenform of eigenvalue δ.

(type ii) For each eigenvalue $\lambda^2 > 0$ of Δ_Γ on $\Omega^0(N; adE|_N)$ we get a pair of eigenvalues $\lambda_\pm = \frac{\delta}{2} \pm \sqrt{\frac{\delta^2}{4} + \lambda^2}$. If $\Delta_\Gamma \phi_\lambda = \lambda^2 \phi_\lambda$ then the corresponding eigenforms of $A_{\Gamma,\delta}$ are

$$\begin{pmatrix} \phi_\lambda \\ -\frac{1}{\lambda_\pm} d_\Gamma \phi_\lambda \end{pmatrix}.$$

(type iii) The spectrum of $- * d_\Gamma|(d^*_\Gamma(\Omega^2(N))$.

LEMMA 8.3.1. *As ω varies over $\mathcal{A}_{\ell,w}(\mathcal{T}_\Gamma, T_0)$ and g varies through B the operators $D_{\omega,\delta}$ are a continuously varying family of Fredholm operators.*

Proof. The continuity assertion is clear. On the cylinder $[0,\infty) \times N$, $D_{\omega,\delta}$ has the form

$$\frac{\partial}{\partial t} + A_{h_\infty,\delta} + o(1)$$

Conjugating $D_{\omega,\delta}$ by the multiplication operator $e^{\frac{-\tau\delta}{2}}$ yields an operator on L^2 forms which agrees with $D_{\omega,\delta} - \frac{\delta}{2}Id$ on $[0,\infty) \times N$. According to Lockhart and McOwen[20], this operator will be Fredholm exactly when $A_{h_\infty,\delta} - \frac{\delta}{2}Id$ is invertible on L^2-forms. Since $\delta/2$ is not in the spectrum of $A_{h_\infty,\delta}$ the lemma is immediate. □

COROLLARY 8.3.2. *For $\omega \in \mathcal{A}_{\ell,w}(\mathcal{T}_\Gamma, T_0)$ the index of $D_{\omega,\delta}$ is independent of ω and of the metric g. If ω is g-ASD then minus the Euler characteristic of $F_\delta(\omega)$ is equal to the index of $D_{\omega,\delta}$.*

Proof. Since the set of metrics and $\mathcal{H}_{out}$ (and hence $\mathcal{A}_{\ell,w}(\mathcal{T}_\Gamma, T_0)$) are connected, indices of this family of operators are all equal. The second statement is clear.

□

Thus, it suffices to compute the index for a single connection $\omega \in \mathcal{A}_{\ell,w}(\mathcal{T}_\Gamma, T_0)$. This we do in the next section.

8.4. Index computations for the fibers

To simplify our computations we shall now assume that G is either $SU(2)$ or $SO(3)$. The main result in this section is the following:

PROPOSITION 8.4.1. *For all $\omega \in \mathcal{M}_{\ell,w}(\mathcal{T}_\Gamma, T_0, g)$ the deformation complex $F_\delta(\omega)$ is Fredholm. The formal dimension is given by*

$$8\ell - \frac{3}{2}(\chi(X) + \sigma(X)) - \frac{h^1_\Gamma - h^0_\Gamma}{2} + \frac{\rho(\Gamma)}{2}$$

where h^i_Γ is the dimension of $H^i(N; ad\Gamma)$ and where $\rho(\Gamma)$ is the Atiyah-Patodi-Singer ρ-invariant for the signature complex twisted by $ad(\Gamma) \otimes C$.

Proof. By Corollary 8.3.2 minus the Euler characteristic of the complex $F_\delta(\omega)$ is independent of $\omega \in \mathcal{M}_{\ell,w}(\mathcal{T}_\Gamma, T_0, g)$ and in fact is equal to the index of $D_{\omega',\delta}$ for any $\omega' \in \mathcal{A}_{\ell,w}(\mathcal{T}_\Gamma, T_0)$. We choose ω' so that in $[-1, \infty) \times N$ it is given by the pullback $\pi^*\Gamma$.

For any $t \geq -1$ we denote $X_t = X - \big((t, \infty) \times N\big)$.

Next we invoke the Atiyah-Patodi-Singer index theorem [2] in order to calculate the index of $D_{\omega',\delta}$. In order to discuss their results in our situation we make the following definitions. Denote by A_Γ the operator $A_{\Gamma,0}$ from the previous section. Let

$$\Pi^+_\Gamma : L^2(\Lambda^0 \oplus \Lambda^1(T^*N)) \to L^2(\Lambda^0 \oplus \Lambda^1(T^*N))$$

be the L^2 projection onto the span of the eigenvectors corresponding to positive eigenvalues of A_Γ, and Π^-_Γ the projection onto the span of the negative eigenvectors. Let $\Omega^1_{\Pi^+_\Gamma}(X_1) = \{\theta \in \Omega^1(X_1) | \Pi^+_\Gamma\theta = 0\}$. The index theorem of APS states that the operator

$$D_{\omega'} = (d^*_{\omega'}, P_+ d_{\omega'}) \colon \Omega^1_{\Pi^+_\Gamma}(X_1) \to \Omega^0(X_1) \oplus \Omega^2_+(X_1),$$

is Fredholm and its index is given by

$$index(D_{\omega'}) = 8c_2(\omega') - \frac{3}{2}(\chi(X) + \sigma(X)) - \frac{h^0_\Gamma + h^1_\Gamma}{2} + \frac{\rho(A_\Gamma)}{2}$$

where h^i_Γ is as before and where $\rho(A_\Gamma)$ is the APS ρ-invariant of the operator A_Γ. (Note we have used a different convention in computing the boundary operator hence we have a different sign in the index theorem for contribution of the ρ invariant.) The operator A_Γ is given by

$$A_\Gamma = \begin{pmatrix} 0 & -d^*_\Gamma \\ -d_\Gamma & *d_\Gamma \end{pmatrix}. \tag{81}$$

This is the boundary operator for the signature complex. Thus:

$$index(D_{\omega'}) = 8c_2(\omega) - \frac{3}{2}(\chi(X) + \sigma(X)) - \frac{h^0_\Gamma + h^1_\Gamma}{2} + \frac{\rho(\Gamma)}{2} \tag{82}$$

where $\rho(\Gamma)$ is the ρ invariant for the signature operator twisted by $ad\Gamma \otimes C$.

PROPOSITION 8.4.2. *For $0 < \delta < 2\mu_\Gamma$ we have*

$$index(D_{\omega'}) + dim\big(H^0(N; ad\Gamma)\big) = index(D_{\omega',\delta}).$$

Proof. We need to introduce the following pair of operators. Let

$$W_\delta = \{\theta \in \Omega^1_{2,\delta}([-1,\infty) \times N) \big| (1 - \Pi^+_\Gamma)\theta|_{\{-1\}\times N} = 0\}$$

and

$$W = \{\theta \in \Omega^1([-1,1] \times N) \big| (1 - \Pi^+_\Gamma)\theta|_{\{-1\}\times N} = 0 \text{ and } \Pi^+_\Gamma\theta|_{\{1\}\times N} = 0\}.$$

We set

$$C_\delta = (e^{-\delta\tau} d^*_\Gamma e^{\delta\tau}, P_+ d_\Gamma)\colon W_\delta \to \Omega^0_{1,\delta}([-1,\infty) \times N) \oplus \Omega^2_{+,\delta}([-1,\infty) \times N)$$

and

$$C = (d^*_\Gamma, P_+ d_\Gamma)\colon W \to \Omega^0([-1,1] \times N) \oplus \Omega^2_+([-1,1] \times N).$$

As above we find these operators to be Fredholm for $0 < \delta < 2\mu_\Gamma$. Since $\tau = 0$ on X_0 it follows that $D_{\omega'}|X_0 = D_{\omega',\delta}|X_0$. It is also clear from the definitions that $C = D_{\omega'}|(X_0 - intX_{-1})$ and $C_\delta = D_{\omega',\delta}|(X - intX_{-1})$. By excision properties of elliptic operators [5] we have that

$$ind(C_\delta) - ind(C) = ind(D_{\omega',\delta}) - ind(D_{\omega'}).$$

It is straightforward to see that

$$\begin{aligned}\dim(\mathrm{Ker}(C_\delta)) &= \dim(H^0(N; ad\Gamma))\\ \dim(\mathrm{Coker}(C_\delta)) &= 0\\ \dim(\mathrm{Ker}(C)) &= \dim(\mathrm{Coker}(C)).\end{aligned}$$

□

Proposition 8.4.2, and the above formulae together complete the proof of Proposition 8.4.1. □

Lemma 8.4.1 shows that the moduli space $\mathcal{M}_{\ell,w}(h, \mathcal{T}_\Gamma, T_0, g)$ has a Fredholm deformation theory and hence is a reasonable space if $\delta/2 \notin Spec(- * d_{\Gamma'}|(Im(d^*_{\Gamma'})$. As δ increases so that $\delta/2$ passes through an eigenvalue of $-*d_{\Gamma'}|(Im(d^*_{\Gamma'}))$ the formal dimension of $\mathcal{M}_{\ell,w}(h, \mathcal{T}_\Gamma, T_0, g)$ decreases by the multiplicity of this eigenvalue. Indeed for a generic metric if $0 < \delta' < \delta$ with $\delta'/2, \delta/2 \notin Spec(- * d_{\Gamma'}|Im(d^*_{\Gamma'}))$ then $\mathcal{M}_{\ell,w}(h, \mathcal{T}_\Gamma, T_0, g)$ is a smooth submanifold of $\mathcal{M}_{\ell,w}(h, \mathcal{T}_\Gamma, T_0, g)$ of codimension equal to the number of eigenvalues of $- * d_{\Gamma'}|(Im(d^*_{\Gamma'})$ between $\delta'/2$ and $\delta/2$. The space that is important when we center things at Γ is $\mathcal{M}_{\ell,w}(h, \mathcal{T}_\Gamma, T_0, g)$ where δ is close to μ_Γ. This may not be the entire space of ASD connections which exponentially decay to ω_h, but the others are related to different flow lines for Ξ^{tr}_Γ in U_Γ.

8.5. The deformation complex for the thickened moduli space

The Lie algebra of $\mathcal{G}_\delta(\Gamma)$ is

$$T_e\mathcal{G}_\delta(\Gamma) = \{s \in \Omega^0_{3,loc}(X) | d_\omega s \in L^2_{2,\delta}\}$$

Lemma 5.2 of [26] implies that every $s \in T_e\mathcal{G}_\delta(\Gamma)$ has a well defined limiting value in the lie algebra of $Stab(\Gamma)$ and that any s with this limiting value equal to zero is in $L^2_{3,\delta}$. Thus there is an exact sequence

$$0 \to \Omega^0_{3,\delta}(X) \to T_e\mathcal{G}_\delta(\Gamma) \to Lie\, Stab(\Gamma) \to 0.$$

By construction there is also an exact sequence for the tangent space to $\mathcal{A}_{\ell,w}(\mathcal{T}_\Gamma, \mathcal{T}_0)$

$$0 \to \Omega^1_{2,\delta}(X) \to T_\omega \mathcal{A}_{\ell,w}(\mathcal{T}_\Gamma, \mathcal{T}_0) \to T_h \mathcal{H}_\Gamma \to 0$$

Fix $h \in \mathcal{H}_\Gamma$ and $\omega \in \mathcal{A}_{\ell,w}(h, \mathcal{T}_\Gamma, \mathcal{T}_0)$. Define $m_\omega : \mathcal{G}_\delta(\Gamma) \to \mathcal{A}_{\ell,w}(\mathcal{T}_\Gamma, \mathcal{T}_0)$ to be the map $m_\omega(\sigma) = \sigma^*\omega$. Let $D_\omega m_\omega : T_e\mathcal{G}_\delta(\Gamma) \to T_\omega \mathcal{A}_{\ell,w}(\mathcal{T}_\Gamma, \mathcal{T}_0)$ be the differential of the action. Also let $n_h : Stab(\Gamma) \to \mathcal{H}_\Gamma$ given by the equation $n_h(\tau) = \tau \cdot h$ and denote its differential by $D_h n_h$: Lie Stab$(\Gamma) \to T_h\mathcal{H}_\Gamma$. Finally let the differential of the modified ASD equation $s\colon \mathcal{A}_{\ell,w}(\mathcal{T}_\Gamma, \mathcal{T}_0) \to \Omega^2_{+,1,\delta}(X)$ be the map

$$D_\omega s\colon T_\omega \mathcal{A}(\mathcal{T}_\Gamma, \mathcal{T}_0) \to \Omega^2_{+,1,\delta}(X)$$

If ω is in addition g-mASD then these maps fit together to form the deformation complex for the thickened moduli space.

$$E_\delta(\omega) : T_e\mathcal{G}_\delta(\Gamma) \xrightarrow{D_\omega m} T_\omega \mathcal{A}_{\ell,w}(\mathcal{T}_\Gamma, \mathcal{T}_0) \xrightarrow{D_\omega s} \Omega^2_{+,1,\delta}(X).$$

The complex $F_\delta(\omega)$ is a subcomplex of $E_\delta(\omega)$ with finite dimensional quotient complex $B(h)$ given by

$$B(h): \qquad \text{Lie Stab}(\Gamma) \xrightarrow{D_h m} T_h\mathcal{H}_\Gamma \longrightarrow 0.$$

Hence, $E_\delta(\omega)$ is a Fredholm complex and

$$Ind(E_\delta(\omega)) = Ind(F_\delta(\omega)) + h^0_\Gamma - h^1_\Gamma.$$

Combining this with Lemma 8.4.1 proves the following

PROPOSITION 8.5.1. *For any* $[\omega] \in \mathcal{M}_{\ell,w}(\mathcal{T}_\Gamma, \mathcal{T}_0, g)$ *the deformation complex* $E_\delta(\omega)$ *is Fredholm. The formal dimension is*

$$8\ell - \frac{3}{2}(\chi(X) + \sigma(X)) + \frac{h^1_\Gamma - h^0_\Gamma}{2} + \frac{\rho(\Gamma)}{2}$$

where h^i_Γ *and* $\rho(\Gamma)$ *are as in the Proposition* 8.4.1.

It follows that there is a long exact sequence relating the cohomology groups of the complexes F, E, and B.

(83)

$$\begin{array}{ccccccccc} 0 & \to & \text{Lie Stab}(\omega) & \to & \text{Lie Stab}(\Gamma) & \to & H^1(F_\delta(\omega)) & \to & H^1(E_\delta(\omega)) \\ & \to & T_h\mathcal{H}_\Gamma & \to & H^2(F_\delta(\omega)) & \to & H^2(E_\delta(\omega)) & \to & 0 \end{array}$$

This discussion has the following useful generalization. If we fix an embedded $Stab(\Gamma)$ invariant submanifold $\mathcal{S} \subset \mathcal{H}_\Gamma$ then there is a deformation complex for mASD connections in $p^{-1}(\mathcal{S})$ which we call $E_\delta(\omega, \mathcal{S})$. The above arguments show this to be a Fredholm complex with $Ind(E_\delta(\omega, \mathcal{S})) = Ind(F_\delta(\omega) + Dim(\mathcal{S}) - Dim(Stab(\Gamma))$.

8.5.1. The case of partial boundary conditions. There is a variant of the above computations which will be useful later. Suppose that $N = N_1 \coprod N_2$. Let $\Gamma = \Gamma_1 \coprod \Gamma_2$ be a flat connection on N. Then we can consider the deformation complex associated to the map

$$\partial_2 \colon \mathcal{M}_{\ell,w}(\mathcal{T}_\Gamma, T_0, g) \to \chi(N_2).$$

The exact same analysis as above show us that the formal dimension of the fiber complex associated to ∂_2 at $[\omega] \in (\partial)^{-1}(\Gamma_1 \coprod \Gamma_2)$ is

$$8\ell - \frac{3}{2}(\chi(X) + \sigma(X)) + \frac{h^1_{\Gamma_1} - h^0_{\Gamma_1}}{2} - \frac{h^1_{\Gamma_2} - h^0_{\Gamma_2}}{2} + \frac{\rho(\Gamma_1) + \rho(\Gamma_2)}{2}. \tag{84}$$

8.6. The local structure of the L^2 and thickened moduli spaces

Once we have the linear theory in place we can use the Kuranishi method to give a finite dimensional local model for both the thickened and L^2 moduli spaces. It is convenient to work now in the based context. We have the following theorem.

THEOREM 8.6.1. *For any $[\omega, f] \in \mathcal{M}^0_{\ell,w}(\mathcal{T}_\Gamma, T_0, g)$ there is an open subset $U \subset H^1(E_\delta(\omega))$ containing it and $Stab(\omega)$-equivariant maps $\psi \colon U \to H^2(E_\delta(\omega))$ and $\phi \colon U \to \mathcal{A}_{\ell,w}(\mathcal{T}_\Gamma, T_0)$ so that the following hold*

(1) *The induced map* $\overline{\phi}: U \times_{Stab(\omega)} \eta_n \to \mathcal{B}^0_{\ell,w}(\mathcal{T}_\Gamma, T_0)$ *is an* $SO(3)$*-equivariant diffeomorphism onto its image.*

(2) *The map* $\overline{\phi}$ *carries* $\psi^{-1}(0) \times_{Stab(\omega)} \eta_n$ *homeomorphically onto a neighborhood of* $[\omega, f]$ *in* $\mathcal{M}^0_{\ell,w}(\mathcal{T}_\Gamma, T_0, g)$.

Furthermore, letting $\mathcal{S}_\Gamma$ *be the Simon data contained in* $\mathcal{T}_\Gamma$, *if* $[\omega, f] \in \mathcal{M}^0_{\ell,w}(\mathcal{S}_\Gamma, g) \subset \mathcal{M}_{\ell,w}(\mathcal{T}_\Gamma, T_0, g)$ *then* $\overline{\psi}$ *maps* $\psi^{-1}(0) \cap \phi^{-1}(Y_\Gamma) \times_{Stab(\omega)} \eta_n$ *homeomorphically onto a neighborhood of* $[\omega, f]$ *in* $\mathcal{M}^0_{\ell,w}(\mathcal{S}_\Gamma, g)$.

8.7. The cohomology groups of the fiber and base complex when ω is flat

When ω is a flat connection the cohomology groups of the $E_\delta(\omega)$ and $F_\delta(\omega)$ are isomorphic to topological invariants of the flat structure. The purpose of this section is to spell out these isomorphisms. These results are closely related to results of [2]. Various special cases are proved in [26],[21],[24], and [23]. The argument here is modeled on that of [26]. To state the results we need to introduce the following notation. X has a natural compactification $\overline{X}$ obtained by adding a copy of N at infinity using the cylindrical structure. The flat connection ω extends to $\overline{X}$. We denote the cohomology groups by $H^i(\overline{X}; ad\omega)$ and the corresponding relative cohomology groups by $H^i(\overline{X}, N; ad\omega)$. Also let $\hat{H}^i(\overline{X}; ad(\omega)) = Im(H^i(\overline{X}, N; ad\omega) \to H^i(\overline{X}; ad\omega))$. By Poincaré duality the paring $H^i(\overline{X}, N; ad) \otimes H^{4-i}(\overline{X}; ad\omega) \to R$ is perfect. In particular the induced pairing $q_\omega : \hat{H}^2(\overline{X}; ad(\omega)) \otimes \hat{H}^2(\overline{X}; ad(\omega)) \to R$ is nondegenerate and symmetric. Let $H^2_+(X; ad\omega)$ (respectively $H^2_-(X; ad\omega)$) denote a maximal positive (respectively negative) definite subspace for q_ω. We denote the dimension of these spaces by $b^2_+(X, \omega)$ and $b^2_-(X, \omega)$ Let $H^2_{\leq 0}(X; ad\omega)$ denote the the annihilator in $H^2(X; ad(\omega))$ of a maximal positive definite subspace in $H^2(X, \partial X; ad(\omega))$. Define $H^2_{\geq 0}(X; ad\omega)$ analogously. We denote the dimensions of these spaces by $b^2_{\geq 0}(X; ad\omega)$ and $b^2_{\leq 0}(X; ad\omega)$ respectively. Viewed as harmonic forms in $\Omega^2_{+,1,\delta}(X)$, $H^2(E_\delta(\omega))$ is a subspace of $H^2(F_\delta(\omega))$ and furthermore the harmonic representatives of $H^2(F_\delta(\omega))$ are self-dual two forms in the kernel of $d_\omega e^{\tau\delta}$. In particular there is a map $j\colon H^2(F_\delta(\omega)) \to H^2(X; ad(\omega))$ given by $j(\omega) = e^{\tau\delta}\omega$.

PROPOSITION 8.7.1. *Let ω be a flat connection on $P \to X$.*

(1) *$H^0(E_\delta(\omega))$ is equal to $LieStab(\omega)$ and $H^0(F_\delta(\omega)) = \{0\}$.*

(2) *$H^1(E_\delta(\omega))$ is isomorphic to $H^1(X; ad(\omega)$.*

(3) *$H^1(F_\delta(\omega))$ is isomorphic to $\hat{H}^1(X; ad(\omega))$.*

(4) *The map j is an injection. $j(H^2(E_\delta(\omega)))$ is contained in $\hat{H}^2(X; ad(\omega))$ and is a maximal positive definite subspace. $j(H^2(F_\delta(\omega)))$ is the annihilator in $H^2(X; ad(\omega))$ of a maximal negative definite subspace.*

Proof. The first item is clear. For θ in either $H^1(F_\delta(\omega))$ or $H^1(E_\delta(\omega))$ it is straightforward to check using Stokes' theorem that $P_+(g)d\theta = 0$ implies that $d\theta = 0$ so there are maps $f\colon H^1(F_\delta(\omega)) \to H^1(X; ad(\omega))$ and $e\colon H^1(E_\delta(\omega)) \to H^1(X; ad(\omega))$. We claim that both maps are injective. Consider the map e first. If $e(\theta) = df$. Now by construction $\theta|_{[T_0,\infty)\times N} = \pi^* h + \alpha$ where $\pi\colon [T_0,\infty) \times N \to N$ is the projection, h is a harmonic one-form on N and $\alpha \in L^2_{2,\delta}$. In particular if θ is exact then $h = 0$. But then $\theta \in L^2_{2,\delta}$ and Lemma 5.2 of [26] implies that f can be chosen to be in $L^2_{3,\delta}$. A similar argument implies injectivity of f. Next we wish to show that both maps have the required range. Consider the range of e first. Fix $[\alpha] \in H^1(X; ad(\omega))$. There is a representative of this class α with the property that $\alpha|_{[T_0,\infty)\times N} = \pi^*(h)$ where h is a harmonic one form. But then $\alpha \in E^1_\delta(\omega)$ so α represents a class β in $H^1(E_\delta(\omega))$ with $j(\beta) = [\alpha]$. The proof that f has the required image is similar. Now consider the last assertion. Fix $w \in H^2(F_\delta(\omega))$ and let $v = e^{\tau\delta}w$. Then v is a closed self dual $L^2_{1,-\delta}$ two-form.

LEMMA 8.7.2. *$v|_{[T_0,\infty)\times N}$ has an asymptotic expansion*

$$v|_{[T_0,\infty)\times N} = P_+(g)(\pi^*(h)) + O(e^{-\mu_\Gamma^- t})$$

where h is a Γ harmonic two form on N.

Proof. As such we can write $v|_{[T_0,\infty)\times N} = dt \wedge \phi_g^{-1} * a + (a)$ where a is an $L^2_{1,\delta}$ section of $\pi^*(\Lambda^2(N)) \to [T_0,\infty) \times N$. The two form a satisfies the differential equation $\frac{\partial a}{\partial t} = d_\Gamma(\phi_g^{-1} * a)$. But $\phi_g = id + O(e^{-\mu t})$ and the largest negative eigenvalue of $d_\Gamma *$ is $-\mu_\Gamma^-$ and so the result follows from standard results for asymptotics of linear equations. □

Thus there is a well defined map $r\colon H^2(F_\delta(\omega)) \to H^2(N; ad(\Gamma))$ which agrees with the composition of j with the restriction $H^2(\overline{X}; ad(\omega)) \to H^2(N; ad(\Gamma))$.

Let us first show that j is injective. Suppose that $j(v) = d_\omega(\alpha)$ where α is a distribution valued one form. First note that we must have $h = 0$. Thus $j(v) \in L^2_{1,\delta}$. It is straight forward to show that we can assume that $\alpha \in L^2_{2,\delta}$. But then consider the integral $\int_X v \wedge *v = \int_X v \wedge v = \int_X v \wedge d_\omega \alpha = 0$ (by Stokes' theorem for forms in $L^2_{1,\delta}$) which implies that $v = 0$. Next we show that $j(H^2(E_\delta(\omega)) \subset \hat{H}^2(\overline{X}, N; ad(\omega))$. It is enough to show that the leading term of the asymptotic expansion for v is zero. To this end we have the following lemma. Let $\delta\colon H^1(N; ad(\Gamma)) \to H^2(F_\delta(\omega))$ be the coboundary map of the long exact sequence 83. $H^1(N; ad(\Gamma))$ is a Hilbert space under the L^2_δ inner product. Thus we may view the adjoint of δ as a map $\delta^*\colon H^2(F_\delta(\omega)) \to H^2(N; \Gamma)$. This map is related to the restriction map by

LEMMA 8.7.3. $\delta^* = \frac{1}{2} r$

Proof. For any $h \in H^1(N;\Gamma)$, $\delta(h)$ is computed by first finding $\theta \in E^1_\delta(\omega)$ so that $\theta|_{[T_0,\infty)\times N} = \pi^*(h) + a$ where $a \in L^2_{2,\delta}$. Then $\delta(h) = [P_+(g) d_\omega(\theta)]$. But then we have for all $w \in H^2(F_\delta(\omega))$ and $h \in H^1(N,\Gamma)$:

$$\begin{aligned} < w, \delta(h) >_{L^2_\delta} &= \int_X w \wedge P_+(g) d_\omega(\theta) e^{\tau\delta} \\ &= \int_X v \wedge P_+(g) d_\omega(\theta) \end{aligned}$$

$$
\begin{aligned}
&= \int_X v \wedge d_\omega \theta \\
&= \lim_{T\to\infty} \int_{\tau \le T} v \wedge d_\omega \theta \\
&= \lim_{T\to\infty} \int_{\tau = T} v \wedge \pi^*(h) + a \\
&= \frac{1}{2} \int_N r(w) \wedge h \\
&= \frac{1}{2} < r(w), h >
\end{aligned}
$$

which proves the desired result. □

One consequence of this lemma is that $H^2(E_\delta(\omega))$ is the kernel of the restriction r and hence $j(H^2(E_\delta(\omega))) \subset \hat{H}^2(X; ad(\omega))$. Next we show that $j(H^2(E_\delta(\omega))$ is a maximal positive definite subspace. Clearly this image is a positive subspace so we need only show that it has maximal dimension. But the index of the complex $E_\delta(\omega)$ is $-\frac{1}{2}(\chi(X; ad(\omega)) + \sigma(X; ad(\omega))) + \frac{h^1_\Gamma - h^0_\Gamma}{2} = b^0(\omega) - b^1(\omega) + b^2_+(\omega)$. Thus the items 1 and 2 we must have $Dim(j(H^2(E_\delta(\omega))) = b^2_+(\omega)$. Hence this subspace is maximal. Lastly we must show that $j(H^2(F_\delta(\omega))$ is the annihilator of a maximal negative definite subspace. To see this note that by the same argument applied to the self-dual deformation complex we can find a maximal negative definite subspace of $\hat{H}^2(X; ad(\omega))$ which is represented by closed ASD L^2_δ two forms. Clearly the pairing with $j(H^2(F_\delta(\omega))$ is zero so we must show that $j(H^2(F_\delta(\omega))$ has the correct dimension. The index of the fiber complex is given by $-\frac{1}{2}(\chi(X; ad(\omega)) + \sigma(X; ad(\omega))) - \frac{h^1_\Gamma - h^0_\Gamma}{2} = Dim(\hat{H}^1(X; ad(\omega)) + b^2_{\ge 0}(\omega)$. The desired result follows directly. □

8.8. On orientations of the moduli spaces when $G = SU(2)$

8.8.1. The group of components of the gauge group and the fundamental group of the quotient. First we determine the group of components of $\mathcal{G}_\delta$ when $G = SU(2)$.

LEMMA 8.8.1. *For the case $G = SU(2)$ there is an exact sequence*

$$H^2(X, \partial X; Z) \xrightarrow{Sq^2} H^4(X, \partial X; Z_2) \to \pi_0(\mathcal{G}_\delta) \to H^3(X, \partial X; Z) \to 0.$$

Proof. The argument is the same as in the closed case. The map to $H^3(X, \partial X; Z)$ is given as follows. Trivialize the bundle away from a point. A gauge transformation σ then determines a map $\hat{\sigma}\colon (X \setminus \{pt\}, \partial X) \to (SU(2), 1)$. Pulling back the fundamental class gives the required map. □

The exact sequence

$$1 \to \mathcal{G}_\delta \to \mathcal{G}_\delta(\Gamma) \to Stab(\Gamma) \to 1$$

gives rise to the long exact sequence

$$\to \pi_1(Stab(\Gamma)) \to \pi_0(\mathcal{G}_\delta) \to \pi_0(\mathcal{G}_\delta(\Gamma)) \to \pi_0(Stab(\Gamma)) \to \{1\}. \tag{85}$$

8.8.2. Orientability of the fibers $\mathcal{M}_{\ell,w}(h, \mathcal{T}_\Gamma, T_0, g)$. By a *formal orientation* of a fiber of Υ, $\mathcal{M}_{\ell,w}(h, \mathcal{T}_\Gamma, T_0, g)$ at a connection ω we mean an orientation of the determinant of the fiber complex $F_\delta(\omega)$. In the case where $\mathcal{M}_{\ell,w}(h, \mathcal{T}_\Gamma, T_0, g)$ is smooth at ω this coincides with a local orientation of this manifold.

By Lemma 8.3.1 $det(Ind)(D_{\omega,\delta})$ is a well defined line bundle over $\mathcal{A}_{\ell,w}(\mathcal{T}_\Gamma, T_0)$. As in the closed case the stabilizers of the connections act trivially on this line bundle so it descends to a line bundle λ over $\mathcal{B}_{\ell,w}(\mathcal{T}_\Gamma, T_0)$. Next we show that λ is trivial and hence that an orientation of the index bundle at any connection ω determines a formal orientation for any $\mathcal{M}_{\ell,w}(\mathcal{T}_\Gamma, T_0)$.

LEMMA 8.8.2. *$\lambda \to \mathcal{M}_{\ell,w}(\mathcal{T}_\Gamma, T_0)$ is trivial.*

Proof. With the exact sequence 85 providing generators for the fundamental group the argument is exactly as in the closed case. □

8.8.3. Orienting the fiber complex $F_\delta(\omega)$. Now we find data which allow us to fix formal orientations of the fiber complex and the complex for the thickened moduli space.

A *homology orientation* for a manifold with cylindrical end X is an orientation of $\hat{H}^1(X) \oplus H^2_{\geq 0}(X) \oplus H^0(\partial X) \oplus H^1(\partial X)$. After tensoring with $su(2)$ the first two groups are the cohomology groups of the fiber complex $F_\delta(\omega_0)$ for a trivial connection ω_0 on X and the second two groups are the kernel of the boundary operator A_{Γ_0} where Γ_0 denotes a trivial flat connection on η.

Fix ω_1 a connection on $R \times \eta$ which agrees with Γ_0 on $(-\infty, -1] \times \eta$ and Γ on $[1, \infty) \times \eta$ and with second Chern number equal to ℓ. For $\delta \leq \frac{1}{2} min\{\mu^{\pm}_{\Gamma}, \mu^{\pm}_{\Gamma_0}\}$ consider the fiber complex $F_\delta(\omega_1)$ associated to ω_1.

LEMMA 8.8.3. *A homology orientation of X and an orientation of the determinant line of $F_\delta(\omega_1)$ determine a formal orientation of the fiber $\mathcal{M}_{\ell'}(h, \mathcal{T}_\Gamma, \mathcal{T}_0)$ for any $\ell' \equiv \ell$ (mod Z)*

Proof. The case for general ℓ' follows from $\ell' = \ell$ by grafting in instantons so we consider the case of $\ell' = \ell$. According to Lemma 8.8.2 it suffices to orient the fiber complex for any connection in $\mathcal{A}_{\ell,w}(\mathcal{T}_\Gamma, \mathcal{T}_0)$. We will work with a connection ω that is trivial on X_{-1} and agrees with ω_1 on $[-1, \infty) \times \eta$. Consider the fiber complex $F_{-\delta}(\pi^*\Gamma_0)$ for the trivial connection $\pi^*\Gamma_0$ on $R \times \eta$ with weight function $e^{-\delta\tau}$. The cohomology groups of this complex are readily computed and are $H^0(F_{-\delta}(\Gamma_0)) = H^0(N; ad\Gamma)$, $H^1(F_{-\delta}(\Gamma_0)) = H^1(N; ad\Gamma)$, and $H^2(F_{-\delta}(\Gamma_0)) = 0$. A straight forward excision argument gives the following equality of determinant lines

$$Det(F_\delta(\omega)) = Det(F_\delta(\omega_0)) \otimes Det(F_{-\delta}(\pi^*(\Gamma_0))) \otimes Det(F_\delta(\omega_1)).$$

But the right hand side of the equation is just

$$Det((\hat{H}^1(X) \oplus H^2_{\geq 0}(X)) \otimes su(2)) \otimes Det((H^0(N; ad\Gamma_0) \oplus H^1(N; ad\Gamma_0)) \otimes su(2)) \otimes Det(F_\delta(\omega_1)).$$

and the result follows immediately. □

8.8.4. Orientation of the thickened moduli space. By a *formal orientation* of the thickened moduli space $\mathcal{M}_{\ell,w}(\mathcal{T}_\Gamma, T_0, g)$ at a connection ω we mean an orientation of the determinant of the Fredholm complex $E_\delta(\omega)$. In the case where $\mathcal{M}_{\ell,w}(\mathcal{T}_\Gamma, T_0, g)$ is smooth at ω this coincides with a local orientation of this manifold.

The long exact sequence 83 implies that a formal orientation of any two of $F_\delta(\omega)$, $E_\delta(\omega)$, and $B(\Gamma)$ determines a formal orientation of the third. This observation together with Lemma 8.8.3 implies the following lemma.

LEMMA 8.8.4. *A homology orientation of X and an orientation of the determinant line of $F_\delta(\omega_1)$ and an orientation of $H^0(N; ad(\Gamma)) \oplus H^1(N; ad(\Gamma))$ determine a formal orientation of the thickened based moduli space $\mathcal{M}^0_{\ell,w}(\mathcal{T}_\Gamma, T_0)$.*

CHAPTER 9

A generic metrics theorem for the thickened moduli space when $G = SU(2)$ or $SO(3)$

Next we consider the local structure of the thickened moduli space for a generic metric. As in the compact case, the arguments only work in the case of $G = SU(2)$ (and $SO(3)$). Given the deformation theory of the previous chapter and the slice theorem which is proved in the next chapter (but which we shall assume here), the techniques of Freed and Uhlenbeck apply with little modification with the usual caveat that these techniques do not apply to the moduli space consisting of flat connections ($\ell = 0$). With this exceptional case in mind we have.

THEOREM 9.0.1. *Let (X, g_0) be a cylindrical end manifold with end modeled on $[0, \infty) \times N$. Let $E \to X$ be a principal $SU(2)$-bundle, and let Γ be a flat connection on $\eta = E|N$. Fix a base point $n \in N$. There is a Baire set of metrics $\widehat{B} \subset B$ so that for any metric $g \in \widehat{B}$ and any $\ell > 0$ the based thickened moduli space $\mathcal{M}_\ell^0(\mathcal{T}_\Gamma, T_0, g)$ is a C^2-manifold of dimension*

$$8\ell - \frac{3}{2}(\chi(X) + \sigma(X)) + \frac{h_\Gamma^1 - h_\Gamma^0}{2} + \frac{\rho(\Gamma)}{2} + 3 \tag{86}$$

where h_Γ^i and $\rho(\Gamma)$ are as in the Proposition 8.4.1. The action of $SO(3)$ is a C^2-action. Furthermore, the map $p^0 \colon \mathcal{M}_\ell^0(\mathcal{T}_\Gamma, T_0, g) \to \mathcal{H}_\Gamma \times_{Stab(\Gamma)} \eta_n$ defined in Chapter 7 is a C^2-map which can be assumed to be transverse to any finite set of smooth submanifolds in $\mathcal{H}_\Gamma \times Stab(\Gamma)\eta_n$. Lastly, if $b_+^2(X) > 0$ we can in addition assume *that the thickened moduli spaces contains no reducible connections and hence is a C^2-manifold of dimension*

$$8\ell - \frac{3}{2}(\chi(X) + \sigma(X)) + \frac{h^1_\Gamma - h^0_\Gamma}{2} + \frac{\rho(\Gamma)}{2}.$$

REMARK 9.0.2. *We have stated these results in the C^2 category. There are analogous results for any C^r, $2 \le r < \infty$. We do not know if these results hold for C^∞, the problem being the lack of a C^∞- center manifold theorem.*

9.1. Perturbations of the based moduli space

As in [12], it is convenient to vary the metric using an automorphism of the tangent bundle. In our context we will need appropriate decay conditions at infinity.

DEFINITION 9.1.1. *Let $\mathcal{C} \subset C^2(Aut(TX))$ consist of those automorphisms φ for which there exists a constant $C > 0$ with $\varphi^*(g_0) \in B$ and with*

$$||\varphi^* g_0 - g_0||_{C^2(\{t\}\times N)} \le Ce^{-\mu t} \tag{87}$$

for all $t > 0$.

Note that this exponential decay is stronger than the decay condition in [26, §6]. $\mathcal{C}$ is topologized as an open subset of the Banach space $C^2_\mu(Aut(T_*X))$, and parameterizes (non-uniquely) an open set of metrics in B. Following [12], let

$$\mathcal{P}\colon \mathcal{C} \times \mathcal{A}_\ell(\mathcal{T}_\Gamma, T_0) \to \Omega^2_{+,\delta}$$

denote the map

$$(\varphi, \omega) \to P_+(\varphi^{-1}F_\omega) - \beta(F_{\hat{\omega}_h})$$

LEMMA 9.1.2. *The map $\mathcal{P}$ is C^1 and for all $h \in \mathcal{H}_\Gamma$ its restriction to $\mathcal{C}\times\mathcal{A}_\ell(h, T_0)$ is C^∞.*

Proof. Because of the decay condition 87, $\varphi \to \varphi^{-1}$ is a smooth map from $\mathcal{C}$ to $\mathcal{C}$. Smoothness in the direction of $\mathcal{A}_\ell(\mathcal{T}_\Gamma, T_0)$ follows from Lemma 7.3.1. □

Note that as a consequence of Lemma 9.1.2, the index of the elliptic complex $E_\delta(\omega)$ is independent of the choice of metric $\varphi^*(g_0)$.

To prove the generic metrics theorem, the key calculation is:

THEOREM 9.1.3. *Suppose that the Chern class $\ell > 0$. Then the restriction of $\mathcal{P}$ to each fiber $\mathcal{C} \times \mathcal{A}_\ell(h, T_0)$ has 0 as a regular value.*

Proof. This is proven in [26], following the lines laid down in [12]. The only point to make here is that since we are working here along the fibers of p the differential of s is $P_+ d_\omega$ and hence the proof of [26] applies. □

Of course the above theorem trivially implies that 0 is a regular value for $\mathcal{P}$. Since by Corollary 10.3.2 the map $\mathcal{P}$ is equivariant with respect to $\mathcal{G}_\delta(\Gamma)$ we also have a result for the quotient.

COROLLARY 9.1.4. *The parameterized thickened based moduli space $\tilde{\mathcal{M}}^0 = \mathcal{P}^{-1}(0) \times_{\mathcal{G}(\Gamma)} \eta_n \subset \mathcal{C} \times \mathcal{B}^0_\ell(\mathcal{T}_\Gamma, T_0)$ is a C^2 infinite dimensional manifold. Furthermore,*

(1) *the projection $\mathcal{C} \times \tilde{\mathcal{M}}^0 \to \mathcal{C}$ is a Fredholm map with index given by Equation 86.*

(2) *The map $p^0 \colon \tilde{\mathcal{M}}^0 \to \mathcal{H}_\Gamma \times_{Stab(\Gamma)} \eta_n$ is a C^2-submersion.*

By Smale's Sard theorem this corollary easily implies all statements in Theorem 9.0.1 except that concerning reducible connections.

REMARK 9.1.5. *The version of unique continuation given in* [10, §4.3] *implies that in this result a generic perturbation of given metric g_0 on an arbitrary open ball in X is a metric for which all the statements of Theorem* 9.0.1 *hold, except possibly the one about reducible connections.*

We turn now to the study of reducible connections in the thickened moduli space.

9.2. Reducible connections

In this section we study when the thickened moduli space contains a reducible connection. As in section 8.7, it is convenient to introduce the natural compactification $\overline{X}$ of X obtained by adding a copy of N at infinity using the cylindrical structure. $\overline{X}$ is a manifold with boundary N. Recall that $\hat{H}^2(X;Z)$ is the image of $H^2(\overline{X},N;Z) \to H^2(\overline{X};Z)$. Analogously, we have $\hat{H}^2(X;R)$ which we also denote by $\hat{H}^2(X)$. There is the cup product pairing

$$q\colon \hat{H}^2(X;Z)\otimes \hat{H}^2(X;Z) \to Z$$

which is nonsingular. Let $C_\ell(X) = \{\alpha \in (\hat{H}^2(X;Z)| < \alpha\cup\alpha, [\overline{X}] >= -\ell\}$.

For $g \in B$ let $H_{\geq 0}(g)$ denote the Harmonic space associated to the second cohomology group of the Fredholm complex

$$0 \to \Omega^0_{3,\delta} \xrightarrow{d} \Omega^1_{2,\delta} \xrightarrow{P_+(g)d} \Omega^2_{+,1,\delta} \to 0.$$

Note that $H_{\geq 0}(g) = \{w \in \Omega^2_{+,1,\delta} | w = *_g w \text{ and } d(e^{\delta\tau}w) = 0\}$. Lemma 8.7.1 implies that the map $j\colon H_{\geq 0}(g) \to H^2(X)$ given by $j(w) = [e^{\tau\delta}w]$ is an isomorphism onto the annihilator of a maximal negative definite subspace of $H^2(\overline{X})$. Let $r^*: H^2(\overline{X}) \to H^2(N)$ denote the restriction.

LEMMA 9.2.1. *$Im(r) = Im(r\circ j)$.*

Proof. This follows from the exact sequence 83 and Lemma 8.7.1. □

Set $H^2_+(g) = Ker\ r\circ j$. Then $H^2_+(g) \subset \hat{H}^2(X)$ is a maximal positive definite subspace.

Just as in the compact case we need to consider the period map $P\colon B \to Gr_{b_+(X)}(\hat{H}^2(X))$ where $P(g) = H_+(g) \subset \hat{H}^2(X)$. The basic result we need is the following analogue of Proposition 4.3.14 of [10].

PROPOSITION 9.2.2. *The period map $P\colon B \to Gr_{b_+(X)}(\hat{H}^2(X))$ is a submersion.*

Proof. The proof of this fact is essentially local via the unique continuation theorem so the argument of [10] applies directly. □

For $\alpha \in C_\ell(X)$ define $W_\alpha \subset Gr_{b^2_+(X)}(\hat{H}^2(X))$ to be the set of planes annihilated by α. The collection $\cup_{\alpha \in C_\ell} W_\alpha$ is a locally finite union of codimension $b^2_+(X)$ submanifolds of $Gr_{b^2_+(X)}(\hat{H}^2(X))$. Set $\mathcal{W}_\alpha = P^{-1}(W_\alpha) \subset B$. The previous proposition implies that the $\cup_{\alpha \in C_\ell} \mathcal{W}_\alpha$ is a locally finite union of codimension $b^2_+(X)$ submanifolds.

The following theorem is the natural generalization to the cylindrical end describing the set of metrics $g \in B$ for which the thickened moduli space $\mathcal{M}^0_\ell(\mathcal{T}_\Gamma, T_0, g)$ contains a reducible connection.

THEOREM 9.2.3. *If $\mathcal{M}_\ell(\mathcal{T}_\Gamma, T_0, g)$ contains a reducible connection then $g \in \cup_\alpha \mathcal{W}_\alpha$*

This theorem directly implies the statement in Theorem 9.0.1 regarding reducible connections in the thickened moduli space.

We begin the proof of this theorem by understanding which connections in a center manifold are reducible.

LEMMA 9.2.4. *Let Γ be a C^∞ flat connection and let $\mathcal{H}_\Gamma$ be a C^2 center manifold. If $h \in \mathcal{H}_\Gamma$ is reducible then h is flat.*

Proof. The map $\zeta \colon \mathcal{N} \subset H^1(N; Ad\Gamma) \to H^\perp$ whose graph is the center manifold is $Stab(\Gamma)$ equivariant and is zero on flat connections in $\mathcal{N}$. Thus it suffices to show that if $a \in H^1(N; ad\Gamma)$ is invariant under a positive dimensional subgroup of $Stab(\Gamma)$ then $\Gamma + a$ is flat. We consider the case that Γ is *ad* trivial; the case where is Γ reduces to a $U(1)$ connection is similar. $H^1(N; ad\Gamma) = H^1(N) \otimes su(2)$ where $SU(2)$ acts via conjugation. Clearly the set of elements which are fixed by at least a $U(1)$ subgroup are those corresponding to decomposable elements of the tensor product. It is easy to see that decomposable elements correspond to flat connections. □

Given this lemma, it is easy to show that any reducible connection in $\mathcal{M}_\ell(\mathcal{T}_\Gamma, \mathcal{T}_0, g)$ is obtained from an ASD connection on a line bundle. And so as in the compact case we are left with finding conditions which determine when a line bundle admits a g-ASD connection.

Next we collect a few remarks regarding connections on a $U(1)$ bundle L over X and the extra data that a choice of flat connection on $L' = L|_N$ gives us. If L' admits a flat connection then $c_1(L')$ is in $H^2_{tor}(N; Z)$, the subgroup of torsion elements. Suppose that L' admits a flat connection Γ. This determines a lift $c_1(L;\Gamma)$ of the real first Chern class $c_1^R(L)$ to $H^2(\overline{X}, N; R)$. If we pick a different flat connection Γ' then of course $c_1(L;\Gamma) = c_1(L;\Gamma') + \beta(\gamma)$ for $\gamma \in H^1(N; R)$ given by $\gamma = [\frac{i}{2\pi}(\Gamma - \Gamma')|_N]$ and β the Bockstein.

Next we show that any finite energy g-ASD connection on L has exponential decay.

LEMMA 9.2.5. *Fix L as above and $g \in B$. Let ω be a g-ASD connection on L with $F_\omega \in L^2$. Then for all*

$$\delta < \frac{\mu^-_{triv}}{4}$$

we have $F_\omega \in L^2_\delta$ and there is a flat connection Γ and gauge transformation σ so that

$$\sigma^*\omega - \pi^*\Gamma|_{[0,\infty)\times\eta} \in L^2_{2,\delta}.$$

In particular $c_1(L;\Gamma) \in H^2(\overline{X}, N)$.

Proof. This follows from Theorem 5.2.2 once we make a few observations. First, for a flat connection Γ on a $U(1)$ bundle $Ad\Gamma$ is trivial we have $-*d_\Gamma = -*d_{\Gamma_{triv}}$. Second, the gradient of Chern-Simons in the $U(1)$ case is the linear vector field $\Gamma + a \in \mathcal{S}_\Gamma \mapsto -*d_\Gamma a \in \mathcal{K}_\Gamma$. Thus the center manifold is the harmonic subspace $\Gamma + H^1(N; Ad\Gamma) \subset \mathcal{S}_\Gamma$. The restriction of the gradient to this slice is 0 hence by Theorem 5.2.2 all flow lines decay exponentially with exponent $\mu^-_{triv}/2$ to a flat connection. (In fact the argument of Theorem 5.2.2 in this simpler linear situation would have given us exponential decay with exponent $-\mu^-_{triv}$ but we do not need this here.) □

LEMMA 9.2.6. *Fix a line bundle L so that $c_1^R(L) \in \hat{H}^2(X)$. For any $g \in B$ the line bundle L admits an L^2 g-ASD connection if and only if*

$$q(c_1(L), w) = 0$$

for all $w \in H_+(g)$. Furthermore, if such a flat connection exists it is unique up to gauge transformation and tensoring with a flat connection on a trivial complex line bundle over X.

Proof. Certainly if L admits an L^2 g-ASD connection then its real first Chern class is represented by a g-ASD two-form. In this case the integrand used in defining the pairing vanishes identically. If there are two such connections then the associated curvature forms of these connections must be the same since the harmonic representative for a class in $\hat{H}^2(X)$ is unique. It is clear that two connections on L with the same curvature form agree up to gauge transformation and tensoring with a flat connection on the trivial line bundle.

Now suppose that L is a line bundle with $c_1^R(L) \in \hat{H}^2(X)$ and $q(c_1(L), w) = 0$ for all $w \in H_+(g)$. We shall show that for some choice of flat connection Γ on L' the lift $c_1(L; \Gamma)$ of $c_1^R(L)$ to $H^2(\overline{X}, \partial\overline{X}; R)$ has an ASD L^2-representative.

Fix a flat connection Γ_0 on L'. Consider the class

$$c_1(L, \Gamma_0) \in H^2(\overline{X}, N; R).$$

By our hypotheses, this class vanishes on $H_+(g)$. Thus, pairing with this class induces a homomorphism from $H_{\geq 0}(g)/H_+(g) \to R$. But $H_{\geq 0}(g)/H_+(g)$ is naturally identified with the image of $H^2(\overline{X}; R)$ in $H^2(N; R)$. Thus, there is a class $a \in H^1(N; R)$ which induces the same pairing on this quotient under this identification. There is a one-form θ on X with $\theta|N$ being closed and representing the class a. Since $\theta|N$ is closed, $\Gamma = \Gamma_0 - \theta|N$ is another flat connection on N. Clearly, $c_1(L,\Gamma) = c_1(L,\Gamma_0) - \beta(a) \in H^2(\overline{X}, N; R)$, and this class is orthogonal to $H_{\geq 0}(g)$.

We claim that $c_1(L,\Gamma)$ has a g-ASD representative, and hence that L has a g-ASD connection. Since $c_1(L,\Gamma)$ is orthogonal to $H_{\geq 0}(g)$, this is immediate from the Fredholm alternative. □

REMARK 9.2.7. *Notice that in this part of the argument as well, it suffices to perturb the original metric g_0 generically on an arbitrary open ball in X. Thus, the perturbation of g_0 needed in order for Theorem 9.0.1 to holds can be taken to be a generic perturbation of g_0 on an arbitrary open ball in X.*

9.3. Comparison with ∂^0

As we noted in the definition, the map $p^0\colon \mathcal{M}^0_\ell(\mathcal{T}_\Gamma, T_0, g) \to \mathcal{H}_\Gamma \times_{Stab(\Gamma)} \eta_n$ locally refines $\partial^0\colon \mathcal{M}^0_\ell(X,g) \to \mathcal{R}(N)$ in the sense that there is an open subset of $\mathcal{M}^0_\ell(X,g)$ which is identified with $(p^0)^{-1}$ of $Y_\Gamma \times_{Stab(\Gamma)} \eta_n \subset \mathcal{H}_\Gamma \times_{Stab(\Gamma)} \eta_n$, and on this subset the map ∂^0 is the composition of p^0 with the flow map $Y_\Gamma \times_{Stab(\Gamma)} \eta_n \to \mathcal{R}(N)$. We have established that p^0 is a C^2-map. Unfortunately, the flow map $Y_\Gamma \times_{Stab(\Gamma)} \eta_n \to \mathcal{R}(N)$ is in general only a continuous map. Thus, it is not true that in general ∂^0 has any differentiable properties. There is at least one context however in which the flow map $Y_\Gamma \times_{Stab(\Gamma)} \eta_n \to \mathcal{R}(N)$ and hence ∂^0 do have differentiable properties. That is covered by the next result.

COROLLARY 9.3.1. *With notation and assumptions as above, suppose that Γ represents a point of $\chi(N)$ whose preimage in $\mathcal{R}(N)$ consists of smooth points of $\mathcal{R}(N)$. Then the based thickened moduli space $\mathcal{M}^0_\ell(\mathcal{T}_\Gamma, T_0, g)$ is identified with an open subset of $\mathcal{M}^0_\ell(X,g)$, and using this identification to give $\mathcal{M}^0_\ell(X,g)$ a local C^2 structure the map $\partial^0\colon \mathcal{M}^0_\ell(X,g) \to \mathcal{R}(N)$ is a C^2 map between C^2-manifolds in a neighborhood of the preimage under ∂^0 of the $SO(3)$-orbit in $\mathcal{R}(N)$ above $[\Gamma]$.*

Proof. The flow on $\mathcal{H}_\Gamma$ is trivial and hence $\mathcal{M}^0_\ell(\mathcal{T}_\Gamma, T_0, g)$ is identified with an open subset of $\mathcal{M}^0_\ell(X, g)$. Furthermore, on this open subset the maps p^0 and ∂^0 agree. The result is now immediate from Theorem 9.0.1. □

9.4. The divisors $\mu(\alpha)$ in the thickened moduli space

Suppose that $\mathcal{M}_\ell(\mathcal{T}_\Gamma, T_0, g)$ is a thickened moduli space which is smooth. Let $\Sigma \subset X$ be a smooth surface. Then the restriction map $\mathcal{M}^0_\ell(\mathcal{T}_\Gamma, T_0, g) \to \mathcal{B}^0(\Sigma)$ is a C^2-map. Thus, if $\Sigma \subset X$ satisfies the conclusion of Lemma 7.5.1 then the pullback under the restriction of a generic section of the complex line bundle $\mathcal{L}_\Sigma$ of Lemma 7.5.3 will be a smooth divisor in $\mathcal{M}_\ell(\mathcal{T}_\Gamma, T_0, g)$ representing the restriction to this thickened moduli space of the class $\mu([\Sigma])$.

CHAPTER 10

Slices and the local structure of $\mathcal{B}^0_{\ell,w}(\mathcal{T}_\Gamma, T_0)$

The purpose of this chapter is to establish slice theorems for the spaces $\mathcal{B}^0_{\ell,w}(\mathcal{T}_\Gamma, T_0)$ in order to prove that these spaces are C^2-Hilbert manifolds. It will also follow from this analysis that $\mathcal{B}^0_{\ell,w,\delta}(\mathcal{R}(N))$ is locally isomorphic to a product of the representation variety and a Hilbert space.

10.1. A product structure for $\mathcal{A}_{\ell,w}(\mathcal{T}_\Gamma, T_0)$

Fix thickening data $\mathcal{T}_\Gamma$ and $T_0 \geq 1$. Fix a product structure $\varphi\colon [0,\infty) \times \eta \to E|([0,\infty) \times N)$. Choose a C^∞ connection ω_0 on E which is equal to constant connection Γ on $[T_0,\infty) \times \eta$ and which has

$$\int_X p_1(ad\omega_0) = -4\ell.$$

Choose a partition of unity supported in $[T_0, T_0+1] \times N$. For $h \in \mathcal{H}_\Gamma$, gluing $\hat{\omega}_h = h(t) + \Theta(h(t))$ to ω_0 using this partition of unity defines a $Stab(\Gamma)$-equivariant, embedding i_{Γ,T_0} of $\mathcal{H}_\Gamma$ into the space $\mathcal{A}_{\ell,w}(\mathcal{T}_\Gamma, T_0)$. We denote $i_{\Gamma,T_0}(h)$ by ω_h.

The following is obvious from the definitions.

LEMMA 10.1.1. *The natural map*

$$\iota_{\Gamma,T_0}\colon \mathcal{H}_\Gamma \times \Omega^1_{2,\delta}(X) \to \mathcal{A}_{\ell,w}(\mathcal{T}_\Gamma, T_0)$$

given by $(h,a) \mapsto i_{\Gamma,T_0}(h) + a$ *is a homeomorphism.*

10.2. Basic Results for d_ω and $\Delta_{\omega,\delta}$

Let us turn now to the basic estimates on operators associated to connections in $\mathcal{A}_{\ell,w}(\mathcal{T}_\Gamma, T_0)$ which we shall need. We shall work with the $L^2_{j,\delta}$-norm on the spaces of forms where the derivatives are measured by covariant derivation with respect to ω_0. Our operators are not exponentially decaying to those for ω_0, but they do differ from those of ω_0 by the sum of an exponentially decaying term and a term which is uniformly bounded in C^k. It turns out that the results are analogues of results for δ-decaying connections and the proofs are along the same lines.

Here are the basic estimates that we need.

LEMMA 10.2.1. *Let $\omega \in \mathcal{A}_{\ell,w}(\mathcal{T}_\Gamma, T_0)$.*

(1) *For any $j \leq 3$ exterior covariant differentiation with respect to ω defines continuous operators*

$$d_\omega : \Omega^i_{j,\delta}(X) \to \Omega^{i+1}_{j-1,\delta}(X)$$

and its adjoint

$$e^{-\tau\delta} d^*_\omega e^{\tau\delta} : \Omega^i_{j,\delta}(X) \to \Omega^{i-1}_{j-1,\delta}(X).$$

(2) *With the operator topology, the operators in Part (1) are continuous functions of $\omega \in \mathcal{A}_{\ell,w}(\mathcal{T}_\Gamma, T_0)$.*

(3) *The operator $d_\omega : \Omega^0_{1,\delta}(X) \to \Omega^1_\delta(X)$ has closed range and trivial kernel.*

(4) *Set $\Delta_{\omega,\delta} = e^{-\tau\delta} d^*_\omega e^{\tau\delta} d_\omega : \Omega^3_\delta(X) \to \Omega^1_\delta(X)$. Then $\Delta_{\omega,\delta}$ is an isomorphism.*

Proof. For any $\omega \in \mathcal{A}_{\ell,w}(\mathcal{T}_\Gamma, T_0)$ there is $h \in \mathcal{H}_\Gamma$ such that we have $\omega = \omega_h + a$ with $a \in \Omega^1_{2,\delta}(X)$. Of course, by definition ω_h is a C^3 connection bounded in the uniform C^3-norm defined by ω_0. It is clear from this that for any $\omega \in \mathcal{A}_{\ell,w}(\mathcal{T}_\Gamma, T_0)$ and any $j \leq 3$, the operators $d_\omega : \Omega^i_{j,\delta}(X) \to \Omega^{i+1}_{j-1,\delta}(X)$ and $e^{-\tau\delta} d^*_\omega e^{\tau\delta}$ are defined and continuous.

For any $h \in \mathcal{H}_\Gamma$ and any $\epsilon > 0$ there is a neighborhood $V \subset \mathcal{H}_\Gamma$ of h such that if $h' \in V$ then $\omega_{h'} = \omega_h - \alpha(h', h)$ with

$$||\alpha(h', h)||_{C^3} < \epsilon.$$

Thus, for any $\omega \in \mathcal{A}_{\ell,w}(\mathcal{T}_\Gamma, T_0)$ and for any $\epsilon > 0$, there is a neighborhood of ω such that for each ω' in this neighborhood we have $\omega' - \omega = a_0 + a_1$ where a_0 is C^3, $a_1 \in \Omega^1_{2,\delta}(X)$, $||a_1||_{L^2_{2,\delta}} < \epsilon$, and the C^3-norm of a_0 is less than ϵ. Since $d_{\omega'} - d_\omega = [(a_0 + a_1), \cdot]$, Part 2 of the lemma follows from the estimates above for a_0 and a_1, using Lemma 7.1.1 in the case of a_1.

The closed range assertion in Part 3 follows as in Lemma 8.3.1. The assertion regarding the kernel follows from the fact that any 0-form in the kernel of d_ω is covariantly constant with respect to ω. But if it is also in $\Omega^0_{1,\delta}(X)$ then it must be zero.

The last statement is an immediate consequence of (1), (2), and (3). □

DEFINITION 10.2.2. *For any $\omega \in \mathcal{A}_{\ell,w}(\mathcal{T}_\Gamma, T_0)$ we define*

$$\mathcal{K}_\omega = Ker(e^{-\tau\delta} d^*_\omega e^{\tau\delta}) \cap \Omega^1_{2,\delta}(X)$$

Here are some consequences of these lemmas.

COROLLARY 10.2.3. *There is a constant $C_1 = C_1(\omega) > 1$ depending continuously on ω such that:*

(1) *For any $\omega \in \mathcal{A}_{\ell,w}(\mathcal{T}_\Gamma, T_0)$, any $i, j \leq 3$, and any $\alpha \in \Omega^i_{j,\delta}(X)$ we have*

$$||d_\omega(\alpha)||_{L^2_{j-1,\delta}} \leq C_1(\omega) \cdot ||\alpha||_{L^2_{j,\delta}} \tag{88}$$

(2) *For all $\omega \in \mathcal{A}_{\ell,\delta}(\mathcal{T}_\Gamma, T_0)$ we have a decomposition*

$$\Omega^1_{2,\delta}(X) = \mathcal{K}_\omega \oplus Im(d_\omega) \cap \Omega^1_{2,\delta}(X). \tag{89}$$

(3) *The following estimate holds for this decomposition: If* $\alpha = k + c$ *with* $k \in \mathcal{K}_\omega$ *and* $c \in Im(d_\omega) \cap \Omega^1_{2,\delta}(X)$*, then*

$$\|\alpha\|_{L^2_{2,\delta}} \geq C_1(\omega)^{-1}\left(\|k\|_{L^2_{2,\delta}} + \|c\|_{L^2_{2,\delta}}\right). \tag{90}$$

(4) *For all* $\alpha \in \Omega^0_{3,\delta}(X)$ *we have*

$$\|d_\omega \alpha\|_{L^2_{2,\delta}} \geq C_1^{-1}(\omega)\|\alpha\|_{L^2_{3,\delta}}.$$

COROLLARY 10.2.4. *There is a continuous function* $\overline{\lambda}_\omega > 1$ *such that for all* $\omega \in \mathcal{A}_{\ell,w}(\mathcal{T}_\Gamma, T_0)$ *and all* $f \in \Omega^0_{2,\delta}(X)$ *we have*

$$\overline{\lambda}_\omega \|f\|_{L^2_{3,\delta}} \geq \|\Delta_{\omega,\delta}(f)\|_{L^2_{1,\delta}} \geq \overline{\lambda}_\omega^{-1}\|f\|_{L^2_{3,\delta}} \tag{91}$$

Proof. This follows directly from the fact that $\Delta_{\omega,\delta}\colon L^2_{3,\delta} \to L^2_{1,\delta}$ is a continuous family of invertible operators. □

10.3. The Slice Theorem and the Topology of the Quotient Space

The estimates on d_ω and $\Delta_{\omega,\delta}$ above lead to a slice theorem for the action of $\mathcal{G}_\delta(\Gamma)$ on $\mathcal{A}_{\ell,w}(\mathcal{T}_\Gamma, T_0) \times_{Stab(\Gamma)} \eta_n$. Here is the main technical assertion used in the proof of the slice theorem.

PROPOSITION 10.3.1. *Let* $\omega = \iota_{\Gamma,T_0}(h_0) + a_0 \in \mathcal{A}_{\ell,w}(\mathcal{T}_\Gamma, T_0)$*. There are continuous positive functions* $\epsilon = \epsilon(\omega)$*, and* $C(\omega)$ *and an open neighborhood* $V_1 \subset \mathcal{H}_\Gamma$ *of* h_0 *such that the following hold. Set* $\epsilon' = \epsilon'(\omega) = C^{-1}(\omega)\epsilon(\omega)$*, and set*

$$S_\omega = \iota_{\Gamma,T_0}\left(\mathcal{H}_\Gamma \times (\{a_0\} + \mathcal{K}_\omega)\right).$$

There is an $\mathcal{G}_\delta$*-invariant neighborhood of* ω *in* S_ω *which is equivariantly homeomorphic to* $U_\omega \times \mathcal{G}_\delta$ *where* U_ω *is an open neighborhood of* ω *in* S_ω *containing* $\iota_{\Gamma,T_0}(V_1 \times (\{a_0\} + B_\epsilon))$ *where* B_ϵ *is the ball of radius* ϵ *in* $\mathcal{K}_\omega$ *centered at* $\{0\}$*. The image of* $U_\omega \times \mathcal{G}_\delta$ *contains* $\iota_{\Gamma,T_0}(V_1 \times (\{a_0\} + D_{\epsilon'}))$ *where* $D_{\epsilon'} \subset \Omega^1_{2,\delta}(X)$ *is the ball of radius* ϵ' *centered at* 0*.*

We call U_ω a slice for the action through ω.

Proof. Set $\beta = \iota_{\Gamma,T_0}(h) - \iota_{\Gamma,T_0}(h_0)$, and set

$$\epsilon(\omega) = \frac{min\{\overline{\lambda}_\omega^{-2}, 1\}}{32C_1(\omega)^3 C}.$$

We choose $V_1 \subset \mathcal{H}_\Gamma$ an open neighborhood such that for all $h \in V_1$ the following estimates hold

$$||e^{-\tau\delta} d^*_\omega e^{\tau\delta} d_{\iota_{\Gamma,T_0}(h)+a_0}(f)||_{L^2_{1,\delta}} \geq \frac{1}{2\overline{\lambda}_\omega} \cdot ||f||_{L^2_{3,\delta}},$$

and

$$||\beta||_{C^2} < \epsilon(\omega).$$

(Here $||\beta||_{C^2}$ refers to the uniform C^2-norm on X.) That such an open subset exists is a consequence of Lemma 10.2.4 and the fact that $d_{\omega'}$ varies continuously with ω'. We set

$$C(\omega) = C_1^2(\omega) max\{8\overline{\lambda}_\omega, 1\}.$$

We shall prove that the proposition holds for these choices.

Clearly, by Lemma 7.1.1, Corollary 10.2.3, and Corollary 10.2.4, both $\epsilon(\omega)$ and $C(\omega)$ are positive, continuous functions of ω. Consider the map

$$\varphi\colon V_1 \times \left(\{a_0\} + \mathcal{K}_\omega\right) \times \mathcal{G}_\delta \to \mathcal{A}_{\ell,w}(\mathcal{T}_\Gamma, T_0)$$

defined by

$$(h, a_0 + a, g) \mapsto \iota_{\Gamma,T_0}(h, a_0 + a) \cdot g.$$

By the parameterized version of the inverse function theorem the proposition will follow directly if we can show that φ is a diffeomorphism and that the operator norm of the differential of φ is uniformly bounded from below by $C(\omega)$.

We begin by proving this lower bound saving the proof of injectivity for later. The differential of this map at $\{h\} \times \{a_0 + a\}$ is identified with

$$j_\omega + d_{\iota_{\Gamma,T_0}(h)+a_0+a} : \mathcal{K}_\omega \oplus \Omega^0_{3,\delta}(X) \to \Omega^1_{2,\delta}(X)$$

where j_ω is the inclusion.

Fix $(b, f) \in \mathcal{K}_\omega \oplus \Omega^0_{3,\delta}(X)$. Our goal is to prove

$$||b + d_{\iota_\Gamma, T_0(h)+a_0+a}(f)||_{L^2_{2,\delta}} \geq C(\omega)^{-1}(||b||_{L^2_{2,\delta}} + ||f||_{L^2_{3,\delta}}) \tag{92}$$

Corollary 10.2.3 allows us to write $b + d_{\iota_\Gamma, T_0(h)+a_0+a} f = k + c$ where $k \in \mathcal{K}_\omega$ and $c \in Im d_\omega \cap \Omega^1_{2,\delta}(X)$. Furthermore, this corollary tells us that

$$||b + d_{\iota_\Gamma, T_0(h)+a_0+a}(f)||_{L^2_{2,\delta}} \geq C_1(\omega)^{-1}\left(||k||_{L^2_{2,\delta}} + ||c||_{L^2_{2,\delta}}\right).$$

Thus to prove inequality 92 is suffices to show

$$||k||_{L^2_{2,\delta}} + ||c||_{L^2_{2,\delta}} \geq C(\omega)^{-1}C_1(\omega)\left(||b||_{L^2_{2,\delta}} + ||f||_{L^2_{3,\delta}}\right). \tag{93}$$

Let us first estimate k. To this end note that we have the following formula for k

$$\begin{aligned} k &= b + d_{\iota_\Gamma, T_0(h)+a_0+a}(f) - d_\omega \Delta^{-1}_{\omega,\delta} e^{-\tau\delta} d^*_\omega e^{\tau\delta} d_{\iota_\Gamma, T_0(h)+a_0+a}(f) \\ &= b + [\beta + a, f] - d_\omega \Delta^{-1}_{\omega,\delta} e^{-\tau\delta} d^*_\omega e^{\tau\delta}[\beta + a, f]. \end{aligned}$$

Now by Lemma 7.1.1

$$||[\beta + a, f]||_{L^2_{2,\delta}} < 2C(||\beta||_{C^2} + ||a||_{L^2_{2,\delta}})||f||_{L^2_{3,\delta}} \leq \frac{1}{8\overline{\lambda}_\omega C_1(\omega)}||f||_{L^2_{3,\delta}}$$

and by Lemmas 7.1.1 and 10.2.3

$$\begin{aligned} ||d_\omega \Delta^{-1}_{\omega,\delta} e^{-\tau\delta} d^*_\omega e^{\tau\delta}[\beta + a, f]||_{L^2_{2,\delta}} &\leq 2CC_1(\omega)^2\overline{\lambda}_\omega(||\beta||_{C^2} + ||a||_{L^2_{2,\delta}})||f||_{L^2_{3,\delta}} \\ &\leq \frac{1}{8\overline{\lambda}_\omega C_1(\omega)}||f||_{L^2_{3,\delta}}. \end{aligned}$$

Thus it follows that for our choice of a and h we have

$$||k||_{L^2_{2,\delta}} > ||b||_{L^2_{2,\delta}} - \frac{1}{4\overline{\lambda}_\omega C_1(\omega)}||f||_{L^2_{3,\delta}}.$$

The next task is to estimate c. Since $b, k \in \mathcal{K}_\omega$ see that

$$e^{-\tau\delta} d^*_\omega e^{\tau\delta}(c) = e^{-\tau\delta} d^*_\omega e^{\tau\delta} d_{\iota_\Gamma, T_0(h)+a_0+a}(f).$$

By assumption there is an $\alpha \in \Omega^0_{3,\delta}(X)$ so that $c = d_\omega\alpha$. Given this Lemma 10.2.3 implies the inequalities

$$\begin{aligned}
||c||_{L^2_{2,\delta}} &= ||d_\omega\alpha||_{L^2_{2,\delta}} \\
&\geq C_1(\omega)^{-1}||\alpha||_{L^2_{3,\delta}} \\
&\geq \overline{\lambda}_\omega^{-1} C_1(\omega)^{-1}||\Delta_{\omega,\delta}\alpha||_{L^2_{1,\delta}} \\
&\geq \overline{\lambda}_\omega^{-1} C_1(\omega)^{-1}||e^{-\tau\delta}d^*_\omega e^{\tau\delta}(c)||_{L^2_{1,\delta}} \\
&\geq \overline{\lambda}_\omega^{-1} C_1(\omega)^{-1}||e^{-\tau\delta}d^*_\omega e^{\tau\delta}d_{\iota_{\Gamma,T_0}(h)+a_0+a}(f)||_{L^2_{1,\delta}}
\end{aligned}$$

Next we examine the last entry in the previous inequality more carefully.

$$e^{-\tau\delta}d^*_\omega e^{\tau\delta}d_{\iota_{\Gamma,T_0}(h)+a_0+a}(f) = e^{-\tau\delta}d^*_\omega e^{\tau\delta}d_{\iota_{\Gamma,T_0}(h)+a_0}(f) + e^{-\tau\delta}d^*_\omega e^{\tau\delta}[a,f].$$

According to the definition of V_1 we have

$$||e^{-\tau\delta}d^*_\omega e^{\tau\delta}d_{\iota_{\Gamma,T_0}(h)+a_0}(f)||_{L^2_{1,\delta}} \geq \overline{\lambda}_\omega^{-1}/2||f||_{L^2_{3,\delta}}.$$

Since $e^{-\tau\delta}d^*_\omega e^{\tau\delta}a = 0$ we see that

$$e^{-\tau\delta}d^*_\omega e^{\tau\delta}[a,f] = - * [*a, d_\omega f].$$

Thus by Lemma 7.1.1 and Corollary 10.2.3,

$$\begin{aligned}
||[*a, d_\omega(f)]||_{L^2_{1,\delta}} &\leq 2C||a||_{L^2_{2,\delta}}||d_\omega(f)||_{L^2_{2,\delta}} \\
&\leq 2CC_1(\omega)||a||_{L^2_{2,\delta}}||f||_{L^2_{3,\delta}} \\
&\leq \frac{1}{8\overline{\lambda}_\omega}||f||_{L^2_{3,\delta}}
\end{aligned}$$

Hence for all f we have

$$||e^{-\tau\delta}d^*_\omega e^{\tau\delta}d_{\iota_{\Gamma,T_0}(h)+a_0+a}(f)||_{L^2_{1,\delta}} > \frac{3}{8\overline{\lambda}_\omega}||f||_{L^2_{3,\delta}}.$$

This proves that

$$||c||_{L^2_{2,\delta}} \geq \frac{3\overline{\lambda}_\omega}{8}||f||_{L^2_{3,\delta}}.$$

We then get the uniform estimate

$$||b + d_{\iota_{\Gamma,T_0}(h)+a_0+a}(f)||_{L^2_{2,\delta}} \geq C_1(\omega)^{-1}(||b||_{L^2_{2,\delta}} + \frac{1}{8\overline{\lambda}_\omega}||f||_{L^2_{3,\delta}}). \tag{94}$$

This last inequality finally establishes inequality 92. Note that this also establishes that the differential is an isomorphism as follows. As $\iota_{\Gamma,T_0}(h)+a_0+a$ varies, this is a continuous family of Fredholm operators. The Fredholm map for $h'=h$ and $a=0$ is is an isomorphism, and hence has index 0. Thus, the index at every point is zero, and hence these linear maps are all onto.

Next let us show that the map

$$V_1 \times \left(\{a_0\}+B_\epsilon\right) \times \mathcal{G}_\delta$$

is globally one-to-one. Fix $h \in V_1$ and $a, b \in B_\epsilon$ and $g \in \mathcal{G}_\delta$ with

$$g^*\left(\iota_{\Gamma,T_0}(h, a_0+a\right) = \iota_{\Gamma,T_0}(h, a_0+b).$$

This means that

$$g^{-1} d_{\iota_{\Gamma,T_0}(h)+a_0}(g) + g^{-1} a g = b$$

or

$$d_{\iota_{\Gamma,T_0}(h)+a_0}(g) = gb - ag$$

and hence

$$e^{-\tau\delta} d_\omega^* e^{\tau\delta} d_{\iota_{\Gamma,T_0}(h)+a_0}(g) = e^{-\tau\delta} d_\omega^* e^{\tau\delta}(gb-ag). \tag{95}$$

This last equation is interpreted in terms of a matrix representation of G. Since g is in $L^2_{3,loc}(X)$ and in the cylinder it is the exponential of an element in $\Omega^0_{3,\delta}(X)$, it follows that the section $h=g-1$ of the vector bundle associated to the matrix representation of G is in $L^2_{3,\delta}$. Using the fact that $e^{-\tau\delta} d_\omega^* e^{\tau\delta} a = e^{-\tau\delta} d_\omega^* e^{\tau\delta} b = 0$ we see that

$$e^{-\tau\delta} d_\omega^* e^{\tau\delta}(gb-ag) = -\{d_\omega g, b\} + \{a, d_\omega g\} = -\{d_\omega h, b\} + \{a, d_\omega h\}$$

where $\{\cdot,\cdot\}$ denotes the pairing obtained from matrix multiplication and pointwise inner product of forms. In particular, both sides of Equation 95 are in $\Omega^0_{1,\delta}(X)$.

From the fact that $a, b \in B_\epsilon$, from Equation 95, Lemma 7.1.1, and Corollary 10.2.3 we see that

$$||e^{-\tau\delta} d_\omega^* e^{\tau\delta}(gb - ag)||_{L^2_{1,\delta}} < ||h||_{L^2_{3,\delta}} \frac{1}{8\overline{\lambda}_\omega}. \tag{96}$$

On the other hand,

$$e^{-\tau\delta} d_\omega^* e^{\tau\delta} d_{\iota_{\Gamma,T_0}(h)+a_0}(g) = e^{-\tau\delta} d_\omega^* e^{\tau\delta} d_{\iota_{\Gamma,T_0}(h)+a_0}(h)$$

and

$$||e^{-\tau\delta} d_\omega^* e^{\tau\delta} d_{\iota_{\Gamma,T_0}(h)+a_0}(h)||_{L^2_{1,\delta}} \geq \frac{1}{2\overline{\lambda}_\omega} ||h||_{L^2_{3,\delta}}. \tag{97}$$

Equations 95, 96, and 97 imply that $h = 0$, and hence that $g = 1$. This proves the one-to-one statement, and completes the proof of the proposition. □

We have the following standard consequence of the slice theorem.

COROLLARY 10.3.2. *The quotient*

$$\mathcal{B}^0_{\ell,w}(\mathcal{T}_\Gamma, T_0) = (\mathcal{A}_{\ell,w}(\mathcal{T}_\Gamma, T_0) \times \eta_n) / \mathcal{G}_\delta(\Gamma)$$

is a Hausdorff space with an G-action which is locally equivariantly homeomorphic to $(\mathcal{H}_\Gamma \times_{Stab(\Gamma)} \eta_n) \times \mathcal{K}_\omega$ *where* $\mathcal{K}_\omega$ *is the Hilbert space*

$$Ker(e^{-\tau\delta} d_\omega^* e^{\tau\delta}) \subset \Omega^1_{2,\delta}(X).$$

Proof. This is immediate from Proposition 10.3.1 and the equation

$$(\mathcal{A}_{\ell,w}(\mathcal{T}_\Gamma, T_0) \times \eta_n)/\mathcal{G}_\delta(\Gamma) = (\mathcal{A}_{\ell,w}(\mathcal{T}_\Gamma, T_0)/\mathcal{G}_\delta) \times_{Stab(\Gamma)} \eta_n.$$

□

This corollary produces local C^2-Hilbert manifold coordinates for $\mathcal{B}^0_{\ell,w}(\mathcal{T}_\Gamma, T_0)$. It is clear that these coordinates have C^2-overlap functions and hence define a C^2-Hilbert manifold structure. Clearly, the G-action is C^2 with respect to this structure.

Similar arguments prove the following.

PROPOSITION 10.3.3. *The space $\mathcal{B}^0_{\ell,w,\delta}(X)$ is locally isomorphic to the product of $\mathcal{R}(N)$ and a Hilbert manifold.*

Part III

Circle Bundles Over Riemann Surfaces

This part is devoted to the study of the case when the end of the cylindrical 4-manifold is a circle bundle over a riemann surface of non-zero Euler characteristic. We derive detailed information about the structure and dimension of the L^2-moduli spaces from the fact that we have detailed information about the nature of the Chern-Simons functional near its critical set in this case. The qualitative nature of the result depends on the sign of the Euler class.

In the last chapter of this part we prove our main theorem about the Donaldson polynomial invariants of 4-manifolds separated by certain nilmanifolds. As a corollary we deduce the vanishing theorem for Donaldson polynomial of 4-manifolds which admit smoothly embedded tori of self-intersection at least 2.

CHAPTER 11

Generalities on Stratified Spaces

Let us begin by recalling the notion of a Whitney stratified subset of a smooth manifold.

DEFINITION 11.0.1. *Let Σ be a topological space. A stratification of Σ is a partition of Σ into a finite number of smooth manifolds*

$$\Sigma = \Sigma_1 \cup \cdots \cup \Sigma_n$$

where for each $i \leq n$ we have $\overline{\Sigma}_i \subset \cup_{j \leq i} \Sigma_j$. The Σ_i are called the strata of the stratification.

DEFINITION 11.0.2. *Let M be a smooth finite dimensional manifold. Let $\Sigma \subset M$ be a closed subset. A Whitney stratification of Σ is a stratification in which each stratum is a smooth submanifold of M and such that the following two conditions (called Whitney Conditions A and B) are satisfied:*

(1) *For any sequence of points $\{y_k\}$ in Σ_j with limit $y \in \Sigma_i$, if the tangent planes $T_{y_k}(\Sigma_j)$ converge to a linear subspace $T \subset T_y M$ then $T_y \Sigma_i \subset T$.*
(2) *If sequences $\{y_k\}$ in Σ_j and $\{y'_k\}$ in Σ_i both have limit $y \in \Sigma_i$ and if the secant lines $\overline{y_k y'_k}$ (with respect to some C^∞ coordinate system on M centered at y) converge to a line ℓ and if the tangent planes $T_{y_k}\Sigma_j$ converge to T then $\ell \subset T$.*

There is a related notion of a smooth stratification with cone bundle neighborhoods. We shall define this notion by defining inductively for each $n \geq 1$ a category LCB(n) of smoothly stratified spaces with local cone bundle neighborhoods with at most n strata. A smoothly stratified space with local cone bundle neighborhoods is simply an object in $LCB = \cup_n LCB(n)$. For each of these categories there will be natural notions of product with a smooth manifold and of the boundary of an object. Also, underlying each object in LCB(n) will be a stratified space with at most n strata. The category LCB(1) is the category of smooth manifolds and smooth maps. Assume that LCB(i) has been defined for all $i < n$. For any closed, finite dimensional object L in LCB(n-1) we denote by $c(L)$ the cone on the topological space underlying L. Its cone point is c, and the open cone is identified with $(0,1] \times L$. An LCB(n) coordinate chart for a stratified space Σ is a homeomorphism ϕ from $U \times c(L)$ to an open subset of Σ where L is a closed, finite dimensional object in LCB(n-1) and U is an open subset in a Hilbert space. We give $U \times c(L)$ a stratification in which $U \times \{c\}$ is the bottom stratum and the others are of the form U times the open cone on strata of L. We require that ϕ be a stratum-preserving map. Coordinate charts $\phi_0 \colon U_0 \times c(L_0) \to \Sigma$ and $\phi_1 : U_1 \times c(L_1) \to \Sigma$ are *compatible* if for each point p in the intersection of their images the following hold.

(1) Near any point $p \in \phi_0(U_0 \times \{c\})$ the composition $\phi_1^{-1} \circ \phi_0|(U \times \{c\})$ is a local diffeomorphism from U_0 to U_1.

(2) If $p \notin \phi_0(U_0 \times \{c\})$ then $\phi_1^{-1} \circ \phi_0$ is an isomorphism of objects in LCB(n-1) near $\phi_0^{-1}(p)$.

(3) Near any point $p \in \phi_0(U_0 \times \{c\})$ for every $r \geq 0$ the composition

$$U_0 \times ((0,t] \times L) \xrightarrow{\phi_1^{-1} \circ \phi_0} U_1 \times c(L_1) \xrightarrow{\pi} U_1$$

converges to the map

$$U_0 \times ((0,t] \times L) \xrightarrow{\pi} U_0 \xrightarrow{\phi_1^{-1} \circ \phi_0} U_1$$

uniformly in the C^r-topology.

A structure in LCB(n) is a stratified space with a maximal atlas of compatible LCB(n-1) coordinate charts covering the space. A morphism in this category is a strata-preserving map f which is C^∞ on each stratum and which with respect to local cone bundle structures for each $r \geq 0$ the maps $\pi' \circ f$ and $(f|U_0) \circ \pi$ given in the diagram

$$\begin{array}{ccc} U_0 \times ((0,t] \times L) & \xrightarrow{f} & U_1 \times ((0,1] \times L') \\ {\scriptstyle\pi}\downarrow & & {\scriptstyle\pi'}\downarrow \\ U_0 & \xrightarrow{f|U_0} & U_1 \end{array}$$

converge uniformly in the C^r-topology as $t \mapsto 0$.

It is well-known that any Whitney stratified subspace of a smooth manifold admits a natural structure of a smoothly stratified space with local cone bundle neighborhoods.

Suppose that Σ and Σ' are smoothly stratified spaces with local cone bundle neighborhoods. A morphism $\phi\colon \Sigma \to \Sigma'$ is said to *pull back the cone bundle neighborhoods* if the local cone bundle neighborhoods in Σ are induced by pullback from the cone bundle neighborhoods in Σ'.

We say that a map from a smoothly stratified space with local cone bundle neighborhoods to a smooth manifold R is transverse to a submanifold $S \subset R$ if the restriction of the map to each stratum is transverse to S. The following is easy to establish.

PROPOSITION 11.0.3. *If $\phi\colon \Sigma \to R$ is transverse to S, then $\phi^{-1}(S)$ inherits the structure of a smoothly stratified space in such a way that the inclusion of $\phi^{-1}(S) \subset \Sigma$ pulls back the cone bundle neighborhoods.*

We have been working with smoothly stratified spaces but in all cases there are the analogous notions of C^r stratified spaces for any $r \geq 1$.

The slice theorem, Corollary 10.3.2 has the following consequence.

PROPOSITION 11.0.4. *For any $\delta > 0$, the space $\mathcal{B}^0_{\ell,w,\delta}(X)$ constructed in Chapter 7 is a smoothly stratified space with local cone bundle neighborhoods and the map* *to $\mathcal{R}(N)$ is a morphism of C^2 stratified spaces which pulls back the local cone bundle neighborhoods. In fact, $\mathcal{B}^0_{\ell,w,\delta}(X)$ is locally isomorphic in the category of smoothly stratified spaces to the product of $\mathcal{R}(N)$ with a Hilbert manifold.*

CHAPTER 12

The relationship of the center manifold to the Kuranishi picture for a general 3-manifold

In general it is hard to get an explicit understanding of the vector field Ξ on the center manifold. In this chapter we compare the vector field of the center manifold with other vector fields on $H^1(N;ad\Gamma)$. In particular we study the Kuranishi deformation picture for the character variety. The Kuranishi picture includes a vector field Z on a neighborhood of 0 in $H^1(N;ad\Gamma)$. We show that this vector field is close to Ξ in a very strong sense – the error term involves the square of the distance from zero times the size of Ξ. (We call vector fields related by such an estimate *quadratically close.*) The vector field Z is much more explicitly given and therefore easier to study. But in general even this vector field is too complicated for direct study. In our examples we find another vector field which is close to Z in the sense described above which is amenable to direct analysis. We then need a method to translate results about this latter vector field back to results about Ξ. This is accomplished in the last section of this chapter where we describe results of this type in the general context of two quadratically close vector fields.

12.1. The Kuranishi Picture

Fix η a principal $SU(2)$ bundle over N. Here we study the way $\chi(N)$ sits inside the space of gauge equivalence classes of connections $\mathcal{B}(N)$. This study of $\chi(N)$

inside $\mathcal{B}(N)$ is achieved via the method of Kuranishi. Recall that for any flat connection Γ we have the slice $S_\Gamma = \{\Gamma\} + Kerd_\Gamma^* \subset \mathcal{A}(\eta)$. Here is the main result.

THEOREM 12.1.1. *Let Γ be a C^∞ flat connection on η. Then there are:*

- *a $stab(\Gamma)$-equivariant neighborhood V_Γ of 0 in $H^1(N; ad\Gamma)$,*
- *a $\mathcal{G}$-invariant neighborhood U of Γ in $\mathcal{A}(\eta)$,*
- *a $stab(\Gamma)$-equivariant real analytic embedding*

$$\varphi_\Gamma \colon V_\Gamma \to S_\Gamma \cap U$$

  whose differential at 0 is the natural inclusion of $H^1(N; ad\Gamma)$ into $Ker(d_\Gamma^*)$,
- *and a $Stab(\Gamma)$-equivariant map*

$$\Phi_\Gamma \colon V_\Gamma \to H^2(N; ad\Gamma)$$

so that φ_Γ is a homeomorphism of $\Phi_\Gamma^{-1}(0)$ onto the space of flat connections in $S_\Gamma \cap U$.

The map Φ_Γ is called the Kuranishi obstruction map. Since we are working on a three manifold, $*\Phi_\Gamma$ is a vector field on $V_\Gamma \subset H^1(N; ad\Gamma)$. So we set

$$Z(h) = - * \Phi_\Gamma(h)$$

(The sign is chosen to match that of ∇f_Γ.) It follows from the theorem that the critical set of Z is identified by φ_Γ with a neighborhood of Γ in the space of flat connections in the slice S_Γ.

The proof of this theorem is essentially standard (as in [12, Lemma 4.7]) except that the equation $F_A = 0$ is not elliptic, even up to gauge equivalence. To recover ellipticity, we show that the curvature equation is (locally) equivalent to an apparently weaker one. Let

$$\Pi \colon \Omega^2_{\frac{1}{2}}(N) \mapsto \Omega^2_{\frac{1}{2}}(N)$$

denote the L^2 orthogonal projection onto $Ker(d_\Gamma) \cap \Omega^2_{\frac{1}{2}}(N)$. We shall show that the equations $F_A = 0$ and $\Pi(F_A) = 0$ are equivalent for connections near to Γ.

LEMMA 12.1.2. *For any $A \in \mathcal{A}(\eta)$ sufficiently close to Γ the restriction of the projection $\Pi|Ker(d_A) \cap \Omega^2_{\frac{1}{2}}(N)$ is one-to-one.*

Proof. Let $A = \Gamma + \alpha$, and fix $a \in Ker(d_A)$ with $\Pi(a) = 0$. First we show that $a \in L^2_{\frac{5}{2}}$. We write $a = d_\Gamma^* u$ for $u \in \Omega^3_{\frac{3}{2}}(N) \cap (Ker d_\Gamma^*)^\perp$. Then

$$d_A d_\Gamma^* u = 0;$$

i.e.,

$$d_\Gamma d_\Gamma^* u = -[\alpha, d_\Gamma^* u].$$

Standard bootstrap arguments show that $a = d_\Gamma^* u \in L^2_{\frac{5}{2}}$.

Now we estimate the $L^2_{\frac{3}{2}}$ norm of a. Since $a \perp Ker(d_\Gamma)$,

$$||a||_{L^2_{\frac{3}{2}}} \leq C||d_\Gamma(a)||_{L^2_{\frac{1}{2}}}$$

for some constant C depending only on Γ. But

$$d_\Gamma(a) = d_A(a) - [\alpha, a] = -[\alpha, a].$$

Hence,

$$||d_\Gamma(a)||_{L^2_{\frac{1}{2}}} \leq C'||\alpha||_{L^2_{\frac{3}{2}}} \cdot ||a||_{L^2_{\frac{3}{2}}}$$

where C' depends only on Γ. If $||\alpha||_{L^2_{\frac{3}{2}}} < (CC')^{-1}$ this implies that $a = 0$.

□

COROLLARY 12.1.3. *There is a neighborhood U_0 of Γ in $\mathcal{A}(\eta)$ so that for all $A \in U_0$ if $\Pi(F_A) = 0$ then $F_A = 0$.*

Proof. This is immediate from the Bianchi identity and the previous lemma. □

Proof of Theorem 12.1.1. We decompose

$$\mathcal{K}_\Gamma = H^1(N; ad(\Gamma)) \oplus d_\Gamma^*\big(\Omega^2_{\frac{5}{2}}(N)\big).$$

Consider the map

$$f\colon H^1(N; ad(\Gamma)) \oplus d_\Gamma^*\big(\Omega^2_{\frac{5}{2}}(N)\big) \mapsto d_\Gamma(\Omega^1_{\frac{3}{2}}(N))$$

given by the formula

$$f(\beta, b) = \Pi'(F_{\Gamma+\beta+b}),$$

where $\Pi'\colon \Omega^2_{\frac{1}{2}} \longrightarrow \Omega^2_{\frac{1}{2}}$ denotes the L^2 projection onto $d_\Gamma(\Omega^1_{\frac{3}{2}})$. This map is real analytic and its differential at $(0,0)$ restricted to the second summand of the domain is

$$b \mapsto d_\Gamma(b)$$

which by the Hodge Theorem is an isomorphism. The implicit function theorem gives a $Stab(\Gamma)$-invariant neighborhood $V_\Gamma \subset H^1(N; ad\Gamma)$ of 0 and a $Stab(\Gamma)$-equivariant map

$$\psi_\Gamma\colon V_\Gamma \longrightarrow d_\Gamma^*\big(\Omega^2_{\frac{5}{2}}(N)\big)$$

satisfying for all $\beta \in V_\Gamma$

$$\Pi'\Big(F_{\Gamma+\beta+\psi_\Gamma(\beta)}\Big) = 0 \tag{98}$$

For all $\beta \in V_\Gamma$ we define

$$\varphi_\Gamma(\beta) = \Gamma + \beta + \psi_\Gamma(\beta).$$

We also define

$$\Phi_\Gamma(\beta) = \Pi\big(F_{\Gamma+\beta+\psi_\Gamma(\beta)}\big).$$

Note that the image of Φ_Γ is contained in $H^2(N; ad\Gamma)$. The implicit function theorem and Corollary 12.1.3 implies the result. □

12.2. The comparison of the vector fields

Recall that the center manifold $\mathcal{H}_\Gamma \subset S_\Gamma$ is given as the graph of a $Stab(\Gamma)$ equivariant C^2 map ζ defined on a neighborhood U of $0 \in H^1(N; ad\Gamma)$. We can arrange that $U = V_\Gamma$ of Theorem 12.1.1. So we pullback Ξ by ζ to get a vector field Ξ' on V_Γ. Clearly then the norms of $\Xi'(h)$ and $\nabla_{\Gamma+h+\zeta(h)} f_\Gamma$ are commensurate. The purpose of this section is to compare the vector fields Z and Ξ'. Of course one thing is obvious, they have the same critical set.

To begin the comparison we note the graphs of both ζ and the Kuranishi map φ_Γ give submanifolds of S_Γ tangent to $\Gamma + H^1(N; ad\Gamma)$ at Γ. The graph of the Kuranishi map is characterized by the property that along it $- * \Pi(F_A)$ is parallel to $H^1(N; ad\Gamma)$ while the center manifold has the property that $\nabla_A f_\Gamma = - * (F_A) + d_A(\Theta(A))$ is tangent along the center manifold. The latter condition is expressed simply, using the notation $\Pi^\perp$ for the projection onto $d^*_\Gamma(\Omega^2_{\frac{5}{2}})$, by:

$$D_h\zeta(\Xi') = \Pi^\perp \left(- * F_{\Gamma+h+\zeta(h)} + d_{\Gamma+h+\zeta(h)}\Theta(\Gamma + h + \zeta(h))\right).$$

A comparison of these properties leads to the following proposition.

PROPOSITION 12.2.1. *There is a $C > 0$ so that for all $h \in V_\Gamma$ we have*

$$|\Xi'(h) - Z(h)| \leq C|h|^2|\Xi'(h)| \tag{99}$$

where the norm is taken with respect to the L^2 norm on harmonic forms.

Proof. By the argument given in the proof Theorem 12.1.1 there is no loss in generality in assuming that there is a constant $C_1 > 0$ and neighborhood $\mathcal{N}$ of $0 \in d^*_\Gamma\Omega^2_{\frac{5}{2}}$ so that for all $\beta \in V_\Gamma$ and $b_0, b_1 \in \mathcal{N}$ we have the inequality

$$C_1||\Pi^\perp(F_{\Gamma+\beta+b_0}) - \Pi^\perp(F_{\Gamma+\beta+b_1})||_{L^2_{\frac{1}{2}}} \geq ||b_0 - b_1||_{L^2_{\frac{3}{2}}} \tag{100}$$

Applying this inequality with $b_0 = \zeta(h)$ and $b_1 = \psi_\Gamma(h)$ and using the characterizing properties of the graphs of these functions we have for all $h \in V_\Gamma$

$$C_1\left(||D_h\zeta(\Xi')||_{L^2_{\frac{1}{2}}} + ||\Pi^\perp d_{\Gamma+h+\zeta(h)}\Theta(\Gamma+h+\zeta(h))||_{L^2_{\frac{1}{2}}}\right) \geq ||\zeta(h)-\psi_\Gamma(h)||_{L^2_{\frac{3}{2}}}.$$

Now consider the function $(\beta, b) \mapsto *\Pi^0(F_{\Gamma+\beta+b})$. The differential of this map at $(0,0)$ is 0. Also we have

$$-*\Pi^0(F_{\Gamma+h+\psi_\Gamma(h)}) = Z(h)$$

and

$$-*\Pi^0(F_{\Gamma+h+\zeta(h)}) = \Xi'(h) + \Pi^0 d_{\Gamma+h+\zeta(h)}\Theta(\Gamma+h+\zeta(h)).$$

From this we have

$$\begin{aligned}
&||\Xi'(h) - Z(h)||_{L^2} \\
&\leq \; C_2||h||_{L^2}||\zeta(h)-\psi_\Gamma(h)||_{L^2_{\frac{3}{2}}} + ||\Pi^0 d_{\Gamma+h+\zeta(h)}\Theta(\Gamma+h+\zeta(h))||_{L^2} \\
&\leq \; C_1C_2||h||\left(||D_h\zeta(\Xi')||_{L^2_{\frac{1}{2}}} + ||\Pi^\perp d_{\Gamma+h+\zeta(h)}\Theta(\Gamma+h+\zeta(h))||_{L^2_{\frac{1}{2}}}\right) \\
&\quad + ||\Pi^0 d_{\Gamma+h+\zeta(h)}\Theta(\Gamma+h+\zeta(h))||_{L^2} \qquad (101)
\end{aligned}$$

Since $D_0\zeta = 0$ there is a constant $C > 0$ so that

$$||D_h\zeta(\Xi')||_{L^2_{\frac{1}{2}}} \leq C||h||_{L^2}||\Xi'(h)||_{L^2}. \qquad (102)$$

To estimate the two other terms we use Inequality 11 on the size of $\Theta(A)$. Applied to the second term, together with the fact that $\Pi^\perp$ is bounded, this gives

$$\begin{aligned}
||\Pi^\perp d_{\Gamma+h+\zeta(h)}\Theta(\Gamma+h+\zeta(h))||_{L^2_{\frac{1}{2}}} &\leq K||\Theta(\Gamma+h+\zeta(h))||_{L^2_{\frac{7}{4}}} \\
&\leq K_0||h+\zeta||_{L^2_{\frac{3}{2}}}||\nabla_{\Gamma+h+\zeta(h)}f_\Gamma||_{L^2} \\
&\leq K_1||h||_{L^2_{\frac{3}{2}}}||\Xi'(h)||_{L^2} \qquad (103)
\end{aligned}$$

Similar for the last term we have

$$\Pi^0 d_{\Gamma+h+\zeta(h)}\Theta(\Gamma+h+\zeta(h)) = \Pi^0([h+\zeta(h), \Theta(\Gamma+h+\zeta(h))]).$$

Again using Inequality 11 to estimate the norm of Θ and the fact that $D_0\zeta = 0$, we find that the last term in Inequality 101 is bounded by

$$C_3||h||_{L^2}^2||\Xi'(h)||_{L^2} \tag{104}$$

Plugging in Inequalities 102,103 and 104 into Inequality 101 gives the desired quadratic bound. □

The conclusion of Theorem 12.2.1 is a useful hypothesis in studying the relationship of the flow lines of a pair of vector fields. We make the following definition.

DEFINITION 12.2.2. *Let Ξ and Z be vector fields defined on an open neighborhood U of the origin in R^n. We say that Ξ and Z are quadratically close if there is a constant $C > 0$ so that for all $x \in U$ we have*

$$|\Xi(x) - Z(x)| \leq C|x|^2|\Xi(x)|. \tag{105}$$

Note that this definition is symmetric in Ξ and Z.

12.3. Comparison of the flows of quadratically close vector fields

PROPOSITION 12.3.1. *Let U be a neighborhood of the origin in a finite dimensional real vector space. Suppose that we have a smooth riemannian metric on U. Let Ξ be a vector field on U and let $f\colon U \to R$ be a real analytic function with $f(0) = 0$. Let $\nabla(f)$ be the gradient of f with respect to the riemannian metric. Suppose that Ξ and $\nabla(f)$ are quadratically close. Then:*

- *If $f(h) \geq 0$ for all $h \in U$, then the stable set for the flow generated by Ξ is the critical set of Ξ.*
- *If $f(h) \leq 0$ for all $h \in U$, then the stable set for the flow generated by Ξ contains a neighborhood of 0.*

Proof. Since f is real analytic, the Lojaszewicz inequality implies that there is a neighborhood $V \subset U$ of the origin and $1/2 \leq \gamma < 1$ such that

$$|f(h)|^{\gamma} \leq |\nabla(f)|.$$

By hypothesis (in either case), the zeros of f in U are all critical points. It follows that the intersection of V with the critical set of f is the intersection of V with the level set $f = 0$. Since $\nabla(f)$ and Ξ are quadratically close, they have the same critical set and Ξ is a gradient-like vector field for f. The case when $f \geq 0$ is now obvious. Now let us suppose that $f \leq 0$.

Since Ξ and $\nabla(f)$ are quadratically close, there is a neighborhood $V' \subset V$ such that for any flow line $A(t)$, $0 \leq t < T$, for Ξ contained in V' we have

$$\left|\frac{dA}{dt} - \nabla_A(f)\right| \leq \frac{1}{2}\left|\nabla_A(f)\right|.$$

Since $f \leq 0$, Simon's length estimate (cf. Theorem 4.2.4) applies in this case to give a constant $\theta > 0$ such that

$$\int_{T_0}^{T_1} \left\|\frac{dA}{dt}\right\| \leq 2/\theta \cdot |f(T_0) - f(T_1)|^{\theta}.$$

Pick $\delta > 0$ such that the closed ball B of radius $(4/\theta)\delta^{\theta}$ centered at the origin is contained in V'. Fix a neighborhood $V'' \subset \frac{1}{2}B$ such that $|f(h)| < \delta$ for all $h \in V''$. We claim that V'' is contained in the stable set of Ξ. Fix $h \in V''$ and let $A(t), 0 \leq t < T$ be the maximal flow line with $A(0) = h$ contained in V'. The length estimate implies that for all $0 \leq t < T$, we have $|h - A(t)| \leq (2/\theta)\delta^{\theta}$ and hence $A(t) \in B \subset V'$. Maximality implies that $T = \infty$. By the length estimate, $A(t)$ has a unique limit point in $V' \cap Crit(f)$. □

REMARK 12.3.2. *We do not need the full strength of quadratic closeness for the conclusion of Proposition12.3.1. The proof shows that the weaker estimate*

$$|\Xi(x) - \nabla_x f| \leq \frac{1}{2}|\nabla_x f|$$

will suffice.

CHAPTER 13

Local data for the representation variety

We now turn to the description of the local structure of the moduli space of finite energy ASD connections on a four manifold with cylindrical end modeled on a circle bundle over a Riemann surface. In view of the analysis of the last chapter, the main problem for us is to understand the stable set of the flow in the center manifold. We do this by finding a sufficiently good approximation to the Kuranishi map. In this way we obtain the results of Taubes in [28].

For the remainder of this chapter we make the following conventions. Let Σ be an oriented Riemann surface of genus h. Let $p\colon N \to \Sigma$ denote a principal circle bundle with $c_1 = k[\Sigma]$ for $k \neq 0$; the orientation of Σ and the fiber orientation give rise to an orientation on N. Fix also a riemannian metric m on Σ of area one. Let θ be any S^1-invariant one form on N which evaluates to 1 on the generator of the circle action. Then $d\theta = -2\pi k p^*(*_m 1)$. We give N the metric in which the kernel of θ is orthogonal to the tangent space to the fiber, the metric on the kernel of θ is pulled up from the base and the length of θ is $\frac{1}{2\pi}$. With respect to this metric, the length of each fiber is 1 and the volume of N is equal to one.

13.1. The fundamental group and its representations into $SU(2)$

Let $G = \pi_1(N)$. A presentation of G is

$$G = < a_1, b_1, \ldots, a_h, b_h, c | c^k = \Pi_{i=1}^h [a_i, b_i], \textit{and } c \textit{ is central} > \tag{106}$$

where h is the genus of Σ.

To begin we recall some facts about the representation variety of $\pi_1(\Sigma)$. Let R_+ denote the space of representations $Hom(\pi_1(\Sigma), SU(2))$. This is a real analytic subvariety of $SU(2)^{2h}$ and hence is a Whitney stratified space. Consequently, it is a smoothly stratified space with local cone bundle neighborhoods.

The top stratum I is the subset of irreducible representations. It is a smooth manifold of dimension $6h-3$. There are two other strata – the $2h+2$-dimensional stratum S consisting of representations with stabilizer isomorphic to S^1, and the 0-dimensional stratum T consisting of ad-trivial representations. Also, let R_- denote the set of ordered $2h$-tuples of matrices

$$\{(A_1, B_1, \ldots, A_h, B_h) \in SU(2)^{2h} | \Pi_{i=1}^{h}[A_i, B_i] = -1\}$$

. It is a smooth manifold of dimension $6h-3$. It is a connected 2^{2h} cover of the component of $Hom(\pi_1(\Sigma); SO(3))$ corresponding to representations with nontrivial w_2.

Let $R = Hom(G, SU(2))$. The next theorem is elementary, and describes R. For t a real number, denote by $e(t)$ the unitary matrix

$$\begin{pmatrix} e^{\imath t} & 0 \\ 0 & e^{-\imath t} \end{pmatrix}$$

THEOREM 13.1.1. *At any irreducible representation R is a smooth variety of dimension $6h-3$. R is a union of components*

$$R = \cup_{0 \leq m < |k|/2} R_m \cup R_{\frac{|k|}{2}}.$$

For any m in the above index set the component R_m is characterized by the property that $r(c)$ is conjugate to $e(2\pi m/k)$ for all $r \in R_m$.

- *R_0 is the pull back of the representation variety of Σ and hence is isomorphic as a stratified space to R_+. We denote the induced strata by I_0, S_0, T_0.*
- *For $1 \leq m < \frac{k}{2}$ the component R_m is $SO(3)$ equivariantly diffeomorphic to $T^{2h} \times S^2$ where $SO(3)$ acts on S^2 by the standard action.*

- *If k is odd $R_{\frac{\lfloor k \rfloor}{2}}$ is diffeomorphic to R_-.*
- *If k is even $R_{\frac{\lfloor k \rfloor}{2}}$ is isomorphic as a stratified space to R_0. In this case we denote the induced strata by $I_{\frac{\lfloor k \rfloor}{2}}, S_{\frac{\lfloor k \rfloor}{2}}, T_{\frac{\lfloor k \rfloor}{2}}$.*

Notice that the only components containing singularities are R_0, and $R_{\frac{\lfloor k \rfloor}{2}}$ if k is even. We denote by $\chi(N) = \bigcup \chi_m(N)$ the induced decomposition of the character variety into components.

13.1.1. The Zariski tangent space. The above result gives complete information about the topology of R, but we need to understand the finer local structure. The ringed structure on $\mathcal{R}(N)$ coming from its embedding in the space of equivalence classes of based connections and the Kuranishi picture is the structure we were describing in the previous section. It follows from work of Goldman and Millson [13, 14] that this ringed structure is identified with its ringed structure as an algebraic set. However, we do not need to make use of such a strong statement, only the more obvious fact that the stratifications and the formal tangent spaces agree. Let us begin with a consideration of the Zariski tangent spaces to the representation variety taking the algebro-geometric point of view now.

In general, let Γ be a finitely presented group. Recall that the space of representations $\mathcal{R}(\Gamma)$ of Γ into $SU(2)$ is naturally the space of real points of an affine algebraic variety defined over Q. A *reducible* representation is one whose image is contained in a circle subgroup of $SU(2)$. There is a closed subvariety $\mathcal{R}_{red}(\Gamma) \subset \mathcal{R}(\Gamma)$ consisting of all reducible representations. There is the the action of $SU(2)$ by conjugation on the representation variety. The Zariski tangent space to $\mathcal{R}(\Gamma)$ at r is the vector space $\mathcal{Z}^1(\Gamma; ad\ r)$ of crossed homomorphisms over r from Γ to $su(2)$. We will denote the Zariski tangent space at r by $\mathcal{T}_r(R(\Gamma))$. The tangent space to the orbit under conjugation consists of the trivial crossed homomorphisms and is identified with $su(2)/su(2)^r$. Hence, the Zariski tangent space to a slice of the conjugation action at a representation r is $H^1(\Gamma; ad\ r)$. The Zariski tangent space to $\mathcal{R}(\Gamma)$ is identified with $H^1(\Gamma; ad\ r) \oplus su(2)/su(2)^r$.

The space $\chi(\Gamma)$ of conjugacy classes of representations Γ into $SU(2)$ is called the *character variety*, even though technically speaking the characters are identified with a semi-algebraic subset of the real points of the algebro-geometry quotient variety.

The theorem of the previous chapter identifies the space of SU(2) representations of $G = \pi_1(N)$; the next lemma computes the cohomology groups and therefore the Zariski tangent spaces to the representation variety $\mathcal{R}(G)$.

LEMMA 13.1.2. *At any irreducible representation r, the Zariski tangent space of $\mathcal{R}(G)$ is of dimension $6h-3$. For components containing reducible representations, the cohomology groups are given as follows:*

(i) *If r is in T_0 or $T_{\frac{\lfloor k \rfloor}{2}}$ (when k is even) then the Zariski tangent space is $6h$-dimensional and is identified with $H^1(N;R) \otimes su(2)$, carrying the obvious action of $SU(2)$.*

(ii) *If r is in S_0 or $S_{\frac{\lfloor k \rfloor}{2}}$ (k even), then $H^1(N; adr)$ splits as a sum $H^1(N; su(2)^r) \oplus H^1(N; C_r)$ where the first vector space is a real vector space of dimension $2h$ and the second is a complex vector space of dimension $2h-2$. At such a representation the Zariski tangent space splits as $R^{2h} \oplus C^{2h-2} \oplus C$ where the action of $stab(r)$ on the first summand is trivial, and on the second and third summands is by complex multiplication.*

(iii) *For $r \in R_m$, $1 \leq m < |k|/2$, the Zariski tangent space at r is equal to $H^1(N; su(2)^r) \oplus su(2)/su(2)^r$ which has dimension $2h+2$.*

In each case, the conjugation actions are the obvious ones.

Proof. This is an easy exercise and is left to the reader. □

13.2. An approximate Kuranishi map

In this section we study the Kuranishi map for a connection $\Gamma \in R_0 \cup R_{\frac{\lfloor k \rfloor}{2}}$. This is done in two steps. First, we give the Kuranishi map explicitly for the flat connections on the Riemann surface Σ. The second step should be to deduce from this the corresponding maps on N. Unfortunately, it does not seem possible to be completely explicit for the flat connections on N. However, Theorem 12.3.1 allows us to deduce the properties of the Kuranishi map, and therefore the flow on the center manifold, by finding a quadratically close approximation to the Kuranishi vector field.

The Kuranishi picture is local so we may pass to $SO(3)$ connections to compute the maps ϕ and Φ. (This may change the stabilizer.) Upon passing to $SO(3)$ connections Γ is pulled back from a flat connection Γ' on Σ. Also note that with the S^1 invariant metric on N the pullback of the harmonic one forms on Σ are harmonic. Thus p induces an isomorphism $p^*: H^1(\Sigma; ad\Gamma') \to H^1(N; ad\Gamma)$. For $\alpha, \alpha' \in H^1(N, ad\Gamma)$ define $[\alpha \wedge \alpha']^0 \in H^0(N; ad\Gamma)$ to be the L^2 projection of $[\alpha \wedge \alpha']$ onto $p^*(H^2(\Sigma; ad\Gamma')$ followed by contraction with $p^*(*1_m)$. The main result is the following.

Proposition 13.2.1. *Let Γ be a reducible connection in $R_0 \cup R_{\frac{|k|}{2}}$ Define $g_\Gamma: H^1(N; ad\Gamma) \to R$ by the equation*

$$g_\Gamma(\alpha) = -\frac{1}{16\pi k} \|[\alpha \wedge \alpha]^0\|^2_{L^2(N)}. \tag{107}$$

The gradient of g_Γ is given by the following equation

$$\nabla_\alpha g_\Gamma = \frac{1}{4\pi k} * ([\alpha \wedge ([\alpha \wedge \alpha]^0 \theta)]). \tag{108}$$

The gradient ∇g_Γ is close to Ξ at $0 \in H^1(N; ad\Gamma)$ in the sense that for some C $|\nabla_\alpha g_\Gamma - \Xi(\alpha)| \le C|\alpha||\nabla_\alpha g_\Gamma|$ for α in a neigborhood of 0 .

Since g_Γ has a definite sign Proposition 12.3.1 and Remark 12.3.2imply the following.

Corollary 13.2.2. *Let Y_Γ denote the stable set of Ξ.*

- *If $k > 0$ then the stable set Y_Γ contains a neighborhood of 0.*

- *If $k < 0$ then there is a neighborhood U of $0 \in H^1(N; ad\Gamma)$ so that the $Y_\Gamma \cap U = Crit(\Xi)$.*

Before getting into the proof of Proposition 13.2.1, let us derive the formula for the gradient of g_Γ. Fix $\alpha, v \in H^1(N; ad\Gamma)$. Then

$$\begin{aligned} d_\alpha g_\Gamma(v) &= \frac{-1}{4\pi k}\int_N < [v \wedge \alpha]^0, ([\alpha \wedge \alpha]^0) > \\ &= \frac{-1}{4\pi k}\int_N tr([v \wedge \alpha][\alpha \wedge \alpha]^0\theta) \\ &= \frac{-1}{4\pi k}\int_N tr(v \wedge [\alpha \wedge [\alpha \wedge \alpha]^0\theta]) \\ &= \int_N tr(v \wedge *(\frac{1}{4\pi k} * ([\alpha \wedge [\alpha \wedge \alpha]^0\theta]))) \end{aligned} \tag{109}$$

This verifies formula 108 for the gradient.

We begin by finding the Kuranishi map for the Riemann surface Σ at a flat connection Γ'.

13.2.1. The Kuranishi picture for a Riemann Surface. Fix an SO(3) bundle $P \to \Sigma$ and a flat connection Γ'. Let $\mathcal{G}_P$ denote the group of L^2_2 gauge transformations of P and $\mathcal{A}_P$ denote the space of L^2_1 connections. As usual $\mathcal{S}_{\Gamma'} = \Gamma' + Ker(d^*_{\Gamma'}) \cap \Omega^1_1(\Sigma, adP)$ denotes the slice to the action of $\mathcal{G}_P$ on $\mathcal{A}_P$ at Γ'. Also, Π^0 will denote the L^2-projection of $kerd^*_\Gamma \cap \Omega^2$ onto $H^2(\Sigma; ad\Gamma')$, and $\Pi^\perp$ the L^2-projection onto $imd_\Gamma \cap \Omega^2$.

PROPOSITION 13.2.3. *Let Γ' be a flat connection on P. Then there are:*

- *a $stab(\Gamma')$ invariant neighborhood V' of 0 in $H^1(\Sigma; ad\Gamma')$*
- *a $\mathcal{G}$ invariant neighborhood U' of Γ' in $\mathcal{A}_P$*
- *a $stab(\Gamma')$ real analytic embedding $\phi' \colon V' \to S_{\Gamma'} \cap U$ whose differential at 0 is the natural inclusion of $H^1(\Sigma)$ into $Ker(d^*_{\Gamma'})$*

Furthermore, the Kuranishi obstruction map $\Phi' \colon V' \to H^2(\Sigma; ad\Gamma')$ is given by $\Phi'(\alpha) = -\frac{1}{2}\Pi^0([\alpha \wedge \alpha])$.

If Γ' is irreducible then Poincaré duality implies that $H^0 = H^2 = \{0\}$ so Γ' is a smooth point of the character variety. As a corollary, we recover in this instance Goldman's theorem [13] that the singularities of the representation variety of a Riemann surface are quadratic.

Proof. The only part of this statement that is not standard is the claim identifying Φ'. Recall that in general the maps ϕ' is characterized by the properties that for all $\alpha \in V'$

(1) $\phi'(\alpha) - \alpha \in Im(d^*_{\Gamma'})$ and

(2) $\Pi^\perp(F_{\Gamma'+\phi'(\alpha)}) = 0.$ (110)

and that $\Phi'(\alpha)$ is defined to be $\Pi^0(F_{\Gamma'+\phi'(\alpha)})$. Set $\phi'(\alpha) - \alpha = \beta$. Then expanding the curvature we have

$$F_{\Gamma'+\phi'(\alpha)} = F_{\Gamma'+\alpha+\beta} \tag{111}$$

$$= d_{\Gamma'}\beta + \frac{1}{2}[\alpha \wedge \alpha] + [\alpha \wedge \beta] + \frac{1}{2}[\beta \wedge \beta] \tag{112}$$

CLAIM 13.2.4. *Let $a \in Im(d^*_{\Gamma'}\Omega^2)$ and $b \in Ker(d^*_{\Gamma'}) \cap \Omega^1$. Then $[a \wedge b]$ is L^2 orthogonal to $H^2(\Sigma; ad\Gamma')$.*

Proof. Write $a = d^*_{\Gamma'}\gamma$. Fix a harmonic representative $\omega \in H^2(\Sigma; ad\Gamma')$. Then

$$\begin{aligned} \int_\Sigma < [d^*_{\Gamma'}\gamma \wedge b], \omega > &= \int_\Sigma < [*d_{\Gamma'} * \gamma \wedge * * b], \omega > \\ &= \int_\Sigma < [d_{\Gamma'} * \gamma \wedge *b], \omega > \\ &= \int_\Sigma d < [*\gamma \wedge *b], \omega > \\ &= 0. \end{aligned}$$

□

It follows that $d_{\Gamma'}\beta + [\alpha \wedge \beta] + \frac{1}{2}[\beta \wedge \beta]$ is L^2 orthogonal to $H^2(\Sigma; ad\Gamma')$ and hence $\Pi^0(F_{\Gamma'+\phi'(\alpha)}) = \frac{1}{2}\Pi^0([\alpha \wedge \alpha]) = \Phi'(\alpha)$. □

We will need some estimates for β. The implicit function theorem implies that β satisfies an estimate of the form $||\beta||_{L_1^2} \leq ||\alpha||_{L^2}^2$. To get a better estimate note the following. Suppose that Γ' is reducible, and fix a nontrivial element σ of the Lie algebra of the stabilizer of Γ'. AdP splits into a real line bundle ξ fixed by σ and its orthogonal complement, $\xi^\perp$. For $a \in \Omega^i(AdP)$ write $a = a^o + a^\perp$ for $a^o \in \Omega^i(\xi)$ and $a^\perp \in \Omega^i(\xi^\perp)$. Then for $a \in \Omega^i(AdP)$ and $b \in \Omega^j(AdP)$ we have $[a^o \wedge b^o] = 0, [a^o \wedge b^\perp], [a^\perp \wedge b^o] \in \Omega^{i+j}(\xi^\perp)$, and $[a^\perp \wedge b^\perp] \in \Omega^{i+j}(\xi^o)$. Finally, note that if $a \in \Omega^0(AdP)$ and $b \in \Omega^i(AdP)$ then we have, for all $n \in N$, the pointwise estimate $|[a^o \wedge b^\perp](n)| \leq |a^o(n)||b^\perp(n)|$. With this said we can state our estimate.

LEMMA 13.2.5. *With notation as above there is a neighborhood U of $0 \in H^1(N; ad\Gamma')$ and a constant $C > 0$ so that for all $\alpha \in U$ the following estimate holds*

$$||\beta|| \leq C||\alpha||||\alpha^\perp|| \tag{113}$$

Proof. Using the characterization of ϕ' (Equation 110) and decomposing Equation 111 along ξ and $\xi^\perp$ gives the following pair of equations.

$$d_{\Gamma'}\beta^o + [\alpha^\perp \wedge \beta^\perp] + \frac{1}{2}[\beta^\perp \wedge \beta^\perp] = -\frac{1}{2}\Pi^\perp([\alpha^\perp \wedge \alpha^\perp]) \tag{114}$$

$$d_{\Gamma'}\beta^\perp + [\alpha^o \wedge \beta^\perp] + [\alpha^\perp \wedge \beta^o] = -\frac{1}{2}\Pi^\perp([\alpha^\perp \wedge \alpha^o]) \tag{115}$$

The implicit function theorem immediately gives the desired inequality. □

13.3. The Kuranishi picture for N

Now we show how to recover the Kuranishi picture for N from that of Σ. Let Γ be a reducible connection in $R_0 \cup R_{\lfloor\frac{k}{2}\rfloor}$ corresponding to the flat $SO(3)$-connection Γ' on $P \to \Sigma$. For $\alpha \in p^*(V')$ set $l(\alpha) = p^*(\phi'(p^*)^{-1}(\alpha)) - \alpha$ where $\phi' \colon V' \to S_{\Gamma'}$ is the Kuranishi embedding. Note that $l(\alpha)$ satisfies an estimate of the form $||l(\alpha)|| \leq C|\alpha|^2$ for some constant $C > 0$.

LEMMA 13.3.1. *Let Γ be a reducible connection $R_0 \cup R_{\lfloor\frac{k}{2}\rfloor}$. Then the following data exist:*

(1) *a $Stab(\Gamma)$ equivariant neighborhood $V_1 \subset p^*(V')$*

(2) *a constant $C > 0$ and*

(3) *a $Stab(\Gamma)$ equivariant function $\psi\colon V \to \mathcal{K}_\Gamma$ satisfying the estimate*

$$\|\psi(\alpha)\| \le C|\alpha| \cdot |\nabla_\alpha g_\Gamma(\alpha)|.$$

so that the Kuranishi embedding $\phi_\Gamma\colon V_1 \to \mathcal{S}_\Gamma$ is given by

$$\phi_\Gamma(\alpha) = \alpha + l(\alpha) + \frac{1}{4\pi k}[\alpha \wedge \alpha]^0\theta + \psi(\alpha).$$

Proof. The curvature of $A = \Gamma + \alpha + l(\alpha) + \frac{1}{4\pi k}[\alpha \wedge \alpha]^0\theta$ satisfies

$$\begin{aligned} F_A &= p^*(F_{\Gamma'+\phi'((p^*)^{-1}(\alpha))}) + d_\Gamma \frac{1}{4\pi k}[\alpha \wedge \alpha]^0\theta \\ &\quad + \frac{1}{4\pi k}[(\alpha + l(\alpha)) \wedge ([\alpha \wedge \alpha]^0\theta)] \qquad (116) \\ &= \frac{1}{2}[\alpha \wedge \alpha]^0 p^*(*_m 1) + \frac{1}{4\pi k}[\alpha \wedge \alpha]^0(-2\pi k p^*(*_m 1)) \\ &\quad + \frac{1}{4\pi k}[(\alpha + l(\alpha)) \wedge ([\alpha \wedge \alpha]^0\theta)] \qquad (117) \\ &= \frac{1}{4\pi k}[(\alpha + l(\alpha)) \wedge [\alpha \wedge \alpha]^0\theta)] \qquad (118) \end{aligned}$$

Then by the defining property of the Kuranishi embedding we have

$$\begin{aligned} 0 &= \Pi^\perp(F_{\Gamma+\phi_\Gamma(\alpha)}) \qquad (119) \\ &= \Pi^\perp\left(\frac{1}{4\pi k}[l(\alpha) \wedge [\alpha \wedge \alpha]^0\theta] + d_\Gamma\psi(\alpha) + [(\alpha + l(\alpha)) \wedge \psi(\alpha)]\right. \\ &\quad \left. + \frac{1}{4\pi k}[\alpha \wedge \alpha]^0\theta \wedge \psi(\alpha) + \frac{1}{2}[\psi(\alpha) \wedge \psi(\alpha)]\right). \qquad (120) \end{aligned}$$

As above $\Pi^\perp$ denotes the L^2 projection on $Im(d_\Gamma) \cap \Omega^2$. The implicit function theorem implies that there is a neighborhood V_2 of $0 \in H^1(N; ad\Gamma)$ so that for all $\alpha \in V_2$ this equation has a unique small solution $\psi(\alpha) \in Im(d_\Gamma^*\Omega^2)$ satisfying the estimate

$$\begin{aligned} ||\psi(\alpha)||_{L^2_{\frac{3}{2}}} &\leq C_1||\Pi(\frac{1}{2}[\alpha + l(\alpha) \wedge [\alpha \wedge \alpha]^0\theta])||_{L^2_{\frac{1}{2}}} \\ &\leq C_2||\alpha + l(\alpha)||_{L^2_{\frac{1}{2}}} ||[\alpha \wedge \alpha]^0\theta]||_{L^2} \end{aligned}$$

If $V_2 \subset U$, where U is the neighborhood of Lemma 13.2.5 then we have

$$\begin{aligned} ||l(\alpha)||_{L^2_{\frac{1}{2}}} ||([\alpha \wedge \alpha]^0\theta)]||_{L^2} &\leq C||\alpha||||\alpha^{\perp}||||[\alpha \wedge \alpha]^0|| \\ &\leq C||\alpha||||[\alpha^{\perp} \wedge [\alpha \wedge \alpha]^0|| \\ &\leq C_3||\alpha||||\nabla_\alpha g_\Gamma||. \end{aligned}$$

This proves the assertion. □

We are now in a position to prove Proposition 13.2.1.

Proof. According Theorem 12.2.1 it is enough to show that $Z = - * F_{\Gamma+\phi(\alpha)}$ is quadratically close to $\nabla_\alpha g_\Gamma$. Using the above lemma we have

$$\begin{aligned} \nabla_\alpha g_\Gamma - Z(\alpha) &= \frac{1}{4\pi k} * ([\alpha \wedge ([\alpha \wedge \alpha]^0\theta)] \\ &\quad - * \Pi\left(\frac{1}{4\pi k}[l(\alpha + l(\alpha)) \wedge [\alpha \wedge \alpha]^0\theta] + [(\alpha + l(\alpha)) \wedge \psi(\alpha)]\right. \\ &\quad \left. + \frac{1}{4\pi k}[\alpha \wedge \alpha]^0\theta \wedge \psi(\alpha) + \frac{1}{2}[\psi(\alpha) \wedge \psi(\alpha)]\right). \end{aligned} \tag{121}$$

From this it is straightforward to obtain the estimate

$$|\nabla_\alpha g_\Gamma - Z(\alpha)| \leq C|\alpha||\nabla_\alpha g_\Gamma|.$$

□

The same circle of of ideas can be used to give an elementary proof of the Lojaszewicz inequality in this case. Again this is only of interest in the case of a reducible connection.

LEMMA 13.3.2. *Let Γ be a reducible connection in $R_0 \cup R_{\frac{|k|}{2}}$. Then there is a constant $C > 0$ and a neighborhood U of $0 \in H^1(N, ad\Gamma)$ so that the following inequality holds for all $\alpha \in U$.*

$$|g_\Gamma(\alpha)|^{\frac{3}{4}} \leq C|\nabla_\alpha g_\Gamma|.$$

REMARK 13.3.3. *Since g_Γ is homogeneous of degree four the exponent in the Lojaszewicz inequality must be greater than or equal to $\frac{3}{4}$.*

Proof. Let U be the neighborhood of Lemma 13.2.5. Note that if $[\alpha \wedge \alpha]^0 = 0$ then the inequality holds. We assume that $[\alpha \wedge \alpha]^0 \neq 0$ and split $ad\eta$ as $\xi \oplus \xi^\perp$, $[\alpha \wedge \alpha]^0$ stabilizes ξ. Then we have

$$\begin{aligned}
|g_\Gamma(\alpha)| &= \frac{1}{16\pi k}||[\alpha \wedge \alpha]^0||^2 \\
&= \frac{1}{16\pi k}||[\alpha^\perp \wedge \alpha^\perp]^0||^2 \\
&= \frac{1}{16\pi k}||[\alpha^\perp \wedge \alpha^\perp]^0||^{\frac{2}{3}}||[\alpha^\perp \wedge \alpha^\perp]^0||^{\frac{4}{3}} \\
&\leq C_1||\alpha^\perp||^{\frac{4}{3}}||[\alpha^\perp \wedge \alpha^\perp]^0||^{\frac{4}{3}} \\
&\leq C_2|\nabla_\alpha g_\Gamma|^{\frac{4}{3}}. \qquad (122)
\end{aligned}$$

The desired inequality follows immediately. □

13.4. Chern-Simons and ρ Invariants

In this section we present a calculation of the Chern-Simons invariants and ρ-invariants of the flat connections on N. These are the main terms which appear in the calculation of the dimension of ASD moduli spaces for manifolds with ends modeled on N. The following table lists the results. At this point we switch point of view and consider $\mathcal{R}(N)$ as the space of flat connections on N.

LEMMA 13.4.1. (1) *For $\Gamma \in R_0$ we have $CS(\Gamma) \equiv 0$ (mod Z) and $\rho(\Gamma) = sign(k)(3 - Dim(Stab(\Gamma))$.*

(2) *For* $\Gamma \in R_m$ *we have* $CS(\Gamma) \equiv -\frac{m^2}{k}$ (mod Z) *and* $\rho(\Gamma) = sign(k)(2 - 8m) + 16m^2/k$.

(3) *For* $\Gamma \in R_{\frac{\lfloor k \rfloor}{2}}$ *we have* $CS(\Gamma) \equiv +\frac{k}{4}$ (mod Z) *and* $\rho(\Gamma) = sign(k)(3 - Dim(Stab(\Gamma))$.

Some parts of the calculations were previously known (cf. [6, 18, 28]). We will take advantage of some basic properties of the Chern-Simons and ρ-invariants.

LEMMA 13.4.2. *Suppose that* Γ_0 *and* Γ_1 *are flat connections on* η *lie in the same component of the space of flat connections. Then* $CS(\Gamma_0) = CS(\Gamma_1)$. *If* Γ_0 *and* Γ_1 *are flat connections on bundles over different* 3*-manifolds* N_0 *and* N_1 *which extend over an oriented cobordism between* N_0 *and* N_1, *then* $CS(\Gamma_0) = CS(\Gamma_1)$.

The proof is a standard manipulation with the Chern-Weil integrand for c_2, and will be omitted. Neither part of this lemma holds for the ρ-invariant; for paths of representations, we have the following weaker statement.

LEMMA 13.4.3. *Suppose that* Γ_t *is a path of flat connections on* η *for which* $h^i(\Gamma_t)$, $i = 0, 1$, *are constant. Then* ρ *is constant along the path* Γ_t.

We remark that if the dimension of the cohomology groups change, then the change in the ρ-invariant is determined by the spectral flow [4].

Proof. (Sketch). Let $W = N \times I$, and let $\Gamma(s)$ ($s \in [0,1]$) be the obvious path of flat connections from Γ_0 to Γ_1, reparameterized to be constant near 0 and 1. Let D_t be the operator $\frac{\partial}{\partial s} + (-d^*_{\Gamma(s)}, *_N d_{\Gamma(s)} - d_{\Gamma(s)})$ acting on $\Omega^0(N; su_2) \oplus \Omega^1(N; su_2)$, with the Atiyah-Patodi-Singer boundary conditions. The hypothesis that the cohomology groups have constant dimension implies that the domains of all of these operators (and their adjoints) may be identified. Under this identification, the family D_t becomes a family of Fredholm operators, which therefore have constant index. On the other hand, the main theorem of [2, 3] identifies this index:

$$index(D_t) = \int_W a(D_t) - \frac{1}{2}\rho(\Gamma_t) + \frac{1}{2}(h^0 + h^1).$$

The integrand $a(D_t)$ is easily computed to be 0; this is essentially the content of the previous lemma. Therefore $\rho(\Gamma_t)$ is constant. □

For flat connections Γ_0, Γ_1 which are cobordant (as in the second part of Lemma 13.4.2), the theorem of [3] says that the difference of the ρ-invariants is given by the signature of the cobordism, with coefficients in the adjoint of the representation which extends over the cobordism.

With these preliminaries in hand, we turn to the actual computations. There are three cases, corresponding to the three types of components of R.

Case I: $\Gamma \in R_0$. These representations are characterized by having $\Gamma(c) = 1$, so they all extend over the disk bundle D_k of Euler class k. Thus $CS \equiv 0 \pmod{Z}$ on R_0, and $\rho(\Gamma)$ may be computed as $3sign(D_k) - sign_{ad\Gamma}(D_k)$. If $ad\Gamma$ is trivial, then $\rho(\Gamma) = 0$. In general, $H^2(D_k; ad\Gamma) \cong R^{\dim stab(\Gamma)}$, and the intersection form is $(\pm)$–definite, depending on the sign of k. So in general,

$$\rho(\Gamma) = \pm(3 - \dim stab(\Gamma))$$

where the sign is the same as that of k.

Case II: $\Gamma \in R_m$, $1 \leq m < \frac{|k|}{2}$. It follows immediately from Theorem 13.1.1 that each component R_m is smooth, so the Chern-Simons and ρ-invariants are determined by their value at a point $\Gamma_m \in R_m$. The Chern-Simons invariant of Γ_m has been computed by Kirk and Klassen [18] and also by Taubes [28] to be $-m^2/k$ (mod Z). For the reader's convenience we give Taubes computation. For the given value of m let Q_m denote the $U(1)$ bundle over Σ with $c_1(Q_m) = m[\Sigma]$. Fix a connection $\tilde{A}$ on Q_m with curvature $F_{\tilde{A}} = -2\pi i m * 1$. Now form the $SU(2)$ bundle P associated to $p^*(Q_m) \rightarrow N$ by the standard representation $U(1) \rightarrow SU(2)$. We give this bundle the connection A induced from $p^*(\tilde{A})$. Note that this connection has

CS invariant 0 (mod Z) since it is pulled up from Σ. Now consider the connection

$$\Gamma_m = A - 2\pi i \frac{m}{k} \theta \begin{bmatrix} 1 & 0 \\ 0 & -1 \end{bmatrix}.$$

Clearly Γ_m is a flat connection corresponding to the component R_m. We compute the relative Chern-Simons invariant between A and Γ_m. The connection A is not flat so the formula is

$$\begin{aligned} CS(\Gamma_m) &\equiv CS(A, \Gamma_m) \\ &\equiv -\frac{1}{8\pi^2} \int_N Tr(a \wedge (2F_A + d_A a + \frac{1}{3} a \wedge a) \\ &= -\frac{1}{8\pi^2} \int_N 2\pi 2i \frac{m}{k} \theta \wedge (2p^*(-2\pi i m * 1) + p^*(2p\pi i m * 1)) \\ &= -\frac{m^2}{k} \int_N \theta \wedge p^*(*1) \\ &= -\frac{m^2}{k} \pmod{Z} \end{aligned} \tag{123}$$

Using the fact that the representation Γ_m has image a finite group, its ρ-invariant may be computed via the G-signature theorem. For a representation β into $U(1)$ sending the generator to $exp(2\pi n i/k)$ the ρ invariant is also called the n-signature, see [7] or [17]. First we consider the case $k > 0$. Then the n-signature is computed in the above references and is $1 - 2n + 2n^2/k$ for $0 < n < k$ and 0 for $n = 0$. The complexified adjoint representation of Γ'_m decomposes as a sum of three such representations with weights $k - 2m, 0, 2m$. Thus we have

$$\rho(\Gamma_m) = 2 - 8m + 16m^2/k \ for \ \ k > 0.$$

The ρ-invariant changes sign when the orientation of the three manifold is changed so we have

$$\rho(\Gamma_m) = -2 + 8m + 16m^2/k \ for \ \ k < 0.$$

Case III: $\Gamma \in R_{\frac{|k|}{2}}$. These representations are characterized by $\Gamma(c) = -1$. Hence $ad\Gamma$ extends over the disk bundle D_k, and the computation of the ρ-invariant is the same as in case I above. So we are left with the computation of the Chern-Simons invariant.

If k is even, then there are *ad*-trivial representations in $R_{\frac{|k|}{2}}$ and the computation is just as in the case of R_m above. The result is $CS(\Gamma) = -k/4$.

If k is odd we must argue differently since in this case there are no reducible connections in the $R_{\frac{|k|}{2}}$. Let A be any connection on D_k which extends Γ. Let $\bar{\Gamma}$ denote the $SO(3)$ connection corresponding to Γ. We can extend $\bar{\Gamma}$ over D_{-k} as a flat $SO(3)$-connection B on a bundle with w_2 Poincaré dual to the zero section. The desired result follows from the following equalities in $\mathbf{R}/\mathbf{Z}$

$$(124) \quad CS(\Gamma) = \int_{D_k} c_2(A) = -\frac{1}{4}\int_{D_k} p_1(\bar{A}) = -\frac{1}{4}\left(\int_{D_k} p_1(\bar{A}) + \int_{D_{-k}} p_1(B)\right)$$
$$= -\frac{1}{4}w_2^2 = k/4.$$

CHAPTER 14

The moduli space

Now we are in a position to give a fairly detailed description of the L^2-moduli space for $SU(2)$-bundles over smooth 4-manifolds with cylindrical ends modeled on a $N \times [0, \infty)$ where N is a circle bundle over a riemannian surface. According to the previous chapter the representation variety $\mathcal{R}(N)$ of representations into $SU(2)$ decomposes as $\coprod R_m$. We denote by $\chi_m(N)$ the component of the character variety which is the image of R_m.

Here is the main result, which gives a global structure to the moduli space.

THEOREM 14.0.1. *Let (X, g_0) be a smooth oriented riemannian 4-manifold with cylindrical end modeled on $[0, \infty) \times N$ where N is orientation-preserving diffeomorphic to a principal S^1-bundle over a riemann surface Σ of genus h with Euler class $k \neq 0$. Let $E \to X$ be a principal $SU(2)$-bundle. Let η be the trivial $SU(2)$-bundle over N. Then for a generic metric $g \in B$ the following hold. For each component R_m of the representation variety $\mathcal{R}(N)$ the moduli space $\mathcal{M}^0_\ell(X, g, R_m)$ of ASD connections asymptotic to R_m is non-empty only if $8\pi^2\ell \equiv CS(R_m) \pmod Z$ and $\ell > 0$. Now fix $\ell > 0$.*

***Case I:** $k > 0$. $\mathcal{M}^0_\ell(X, g, R_m)$ is a smooth manifold with a smooth action of $SO(3)$. Its dimension is given by*

$$8\ell - \frac{3}{2}(\chi(X) + \sigma(X)) + \frac{6h+3}{2} \quad if \quad m = 0, \frac{k}{2}$$

$$8\ell - \frac{3}{2}(\chi(X)+\sigma(X)) - 4m + \frac{8m^2}{k} + h + \frac{7}{2} \quad if \quad 1 \le m < \frac{k}{2}.$$

Let $V_m \subset R_m$ be the open subset of smooth points of R_m. Then the restriction of the boundary mapping

$$\partial^0 : \mathcal{M}^0_\ell(X,g,V_m) \to V_m$$

is a smooth map. Lastly, if $b_2^+(X) > 0$, then there are no reducible connections in $\mathcal{M}_\ell(X,g)$, and hence the moduli spaces $\mathcal{M}_\ell(X,g,\chi_m(N))$ are smooth manifolds whose dimensions are three less that the numbers given in the above equations for the corresponding based moduli spaces. Also, letting V be the open subset of $\chi(N)$ whose preimage in $\mathcal{R}(N)$ is $\coprod_m V_m$, we have that the map $\partial\colon \mathcal{M}_\ell(X,g,V) \to V$ is a smooth map.

Case II: $k < 0$. *For each m, $1 \le m < |k|/2$, the space $\mathcal{M}^0_\ell(X,g,R_m)$ is a smooth manifold with a smooth action of $SO(3)$. Its dimension is*

$$8\ell - \frac{3}{2}(\chi(X)+\sigma(X)) + h + 4m + \frac{8m^2}{k} + \frac{3}{2}.$$

The moduli spaces $\mathcal{M}^0_\ell(X,g,R_0)$ and $\mathcal{M}^0_\ell(X,g,R_{\frac{|k|}{2}})$ are stratified spaces. The preimage under ∂^0 of any stratum of $R_0 \coprod R_{\frac{|k|}{2}}$ is a stratum. Furthermore, ∂^0 is is transverse to the stratification in the sense that the cone bundle neighborhoods for the stratification of R_0 and $R_{\frac{|k|}{2}}$ pull back to cone bundle neighborhoods for the stratification of the based moduli space. Let T, S, I be the strata of $R_0 \coprod R_{\frac{|k|}{2}}$ consisting respectively of the ad-trivial connections, the connections with stabilizer S^1, and the irreducible connections. Notice that unless $k \equiv 0 \pmod 2$ then $T \cap R_{\frac{|k|}{2}} = \emptyset = S \cap R_{\frac{|k|}{2}}$. Then we have:

(1) *$dim(\mathcal{M}^0_\ell(X,g,T)) = 8\ell - \frac{3}{2}(\chi(X)+\sigma(X)) - 3h + \frac{3}{2}$.*
(2) *$dim(\mathcal{M}^0_\ell(X,g,S)) = 8\ell - \frac{3}{2}(\chi(X)+\sigma(X)) - h + \frac{7}{2}$.*
(3) *$dim(\mathcal{M}^0_\ell(X,g,I)) = 8\ell - \frac{3}{2}(\chi(X)+\sigma(X)) + 3h - \frac{3}{2}$.*

If $b_2^+(X) > 0$ then there are no reducible connections in $\mathcal{M}_\ell(X,g)$. Hence, letting $\overline{T}, \overline{S}, \overline{I}$ be the images of T, S, I in $\chi(N)$, the dimensions of $\mathcal{M}_\ell(X,g,\overline{T}), \mathcal{M}_\ell(X,g,\overline{S})$, *and $\mathcal{M}_\ell(X, g, \overline{I})$ are three less than the dimensions given above for the corresponding based moduli spaces.*

It is also convenient to re-write these numbers in terms on the natural completion of X to a closed manifold. Let $\hat{X}$ be the closed 4-manifold obtained from X by gluing in the D^2-bundle over Σ with Euler class $-k$.

COROLLARY 14.0.2. *With X as above and $b_2^+(X) > 0$, the dimension of $\mathcal{M}_\ell(X, g, \chi_m)$ in the case $k > 0$ is equal to*

$$8\ell - \frac{3}{2}(\chi(\hat{X}) + \sigma(\hat{X})) \quad if \quad m = 0, \frac{k}{2}$$

and

$$8\ell - \frac{3}{2}(\chi(\hat{X}) + \sigma(\hat{X})) - 4m + \frac{8m^2}{k} + \chi(\Sigma) \quad if \quad 1 \leq m < \frac{k}{2}.$$

In case $k < 0$ the dimension of $\mathcal{M}_\ell^0(X, g, R_m)$ is equal to

$$8\ell - \frac{3}{2}(\chi(\hat{X}) + \sigma(\hat{X})) + 4m + \frac{8m^2}{k} + \chi(\Sigma) + 1 \quad if \quad 1 \leq m < \frac{k}{2}$$

and the top strata of $\mathcal{M}_\ell(X, g, R_0)$ and $\mathcal{M}_\ell(X, g, R_{\frac{|k|}{2}})$ are both of dimension

$$8\ell - \frac{3}{2}(\chi(\hat{X}) + \sigma(\hat{X})).$$

Notice that for every $k \neq 0$ if we glue in a non-ad-trivial flat connection over the disk bundle there is no obstruction to deforming the result to an ASD connection on the closed manifold $\hat{X}$. (This uses the gluing theorem of [23] and the calculations in section 8.7 to identify the obstruction bundle.) Thus, this procedure gives us a diffeomorphism between $\mathcal{M}_\ell(X, g, \overline{I}_0)$ and an open subset of $\mathcal{M}_\ell(\hat{X}, \hat{g})$ where $\hat{g}$ is an appropriate glued up metric on $\hat{X}$. This shows that in both cases the dimension of the top stratum of $\mathcal{M}_\ell(X, g, R_0)$ has the same dimension as $\mathcal{M}_\ell(\hat{X}, \hat{g})$, which checks with the formulas given above.

REMARK 14.0.3. *It will follow from the proof given here and the remark after the proof of Theorem 9.0.1 that we can take the metric g for which this result holds to* *agree with g_0 outside an arbitrary ball in X. In particular, we can find a metric g as in the theorem which agrees with g_0 on the cylinder $[0,\infty)\times N\subset X$.*

Proof. Let us begin with the case $k>0$. According to Item 1 of Corollary 13.2.2 in this case for any flat connection Γ on N the stable set Y_Γ of the critical set of Ξ is an entire neighborhood of $\{0\}$ in $H^1(N;ad\Gamma)$. This means that for any metric $g\in B$ there is an open subset of the based thickened moduli space centered at Γ which is identified with a neighborhood of $(\partial^0)^{-1}(\Gamma)$ in $\mathcal{M}^0_\ell(X,g)$. We can cover the moduli space $\mathcal{M}^0_\ell(X,g)$ by finitely many neighborhoods each of this type. According to Theorem 9.0.1 for a generic choice of metric in B all these based thickened moduli spaces will be C^2-manifolds. In this way we cover the based L^2-moduli space by C^2-manifold charts. We must study the overlap of these charts.

Let us first consider what happens if we fix Γ but change the center manifold. Let $\mathcal{H}_\Gamma$ and $\mathcal{H}'_\Gamma$ be C^2-center manifolds centered at Γ given by graphs of functions ζ and ζ' over an open neighborhood of $\{0\}$ in $H^1(N;ad\Gamma)$. Since every flow line in $\mathcal{H}'_\Gamma$ converges to the critical set, it corresponds in the usual way to an ASD connection on $[0,\infty)\times\eta$. By the exponential approach result of Theorem 5.2.2 associated to each $h'\in\mathcal{H}'_\Gamma$ there is $h\in\mathcal{H}_\Gamma$ such that the flow lines in the two center manifolds with these initial conditions are exponentially close. This defines a continuous function from $\mathcal{H}'_\Gamma\to\mathcal{H}_\Gamma$ which is clearly a homeomorphism. The point is to show that it is a C^1-diffeomorphism. One way to do this is to apply the exponential approach result to the tangent bundles to center manifolds for the induced flow.

This shows that if we vary the center manifold but not the flat connection then the C^1-structure is unchanged.

Now let us consider two flat connections Γ and Γ'. Because of the above argument, there is no loss of generality in assuming that the center manifold $\mathcal{H}_{\Gamma'}$ for Γ' is contained in the center manifold $\mathcal{H}_\Gamma$ for Γ. We do not assume however that the dimension of the Zariski tangent spaces at Γ and Γ' are the same. If they do have the same dimension, then there is nothing to show. Suppose that they do not. Then $dim\, H^1(N; ad\Gamma') < dim\, H^1(N; ad\Gamma)$. But there are flow lines in $\mathcal{H}_\Gamma$ converging to the critical set in $\mathcal{H}_{\Gamma'}$ which have exponential approach to the critical set with exponent greater than the δ for Γ' but less than the δ for Γ. We consider the decomposition of a neighborhood $\mathcal{N}$ of $\mathcal{H}_{\Gamma'}$ in $\mathcal{H}_\Gamma$ into the center and stable manifolds centered at Γ for the flow. (There is no unstable set since all flow lines converge to the critical set.) The point is that the center-stable decomposition of $\mathcal{N}$ is a C^2-decomposition. From this the C^2-compatibility of the charts is straightforward.

Using this one shows that the C^2-center manifold leads to a C^1-structure on the L^2-moduli space. Similarly, a collection of C^k-center manifolds leads to a C^{k-1}-structure on the L^2-moduli space, compatible with the C^1 structure. Since there is a C^k center manifold for every $1 \leq k < \infty$, this gives the L^2-moduli space a C^∞-structure. Clearly, in light of this and Corollary 9.3.1, the map ∂^0 is a smooth map.

The dimensions of $\mathcal{M}^0_\ell(X, g, R_m)$ are easily computed, given this identification, the dimension result of Theorem 9.0.1 and the computations in Lemma 13.4.1.

Now let us consider the case when $k < 0$. According to Corollary 13.2.2 for any flat connection Γ on η the stable set in the center manifold $\mathcal{H}_\Gamma$ is exactly the intersection of $\mathcal{H}_\Gamma$ with the subset of flat connections. Thus, there is $\delta > 0$ depending on Γ such that for any metric $g \in B$ every g-ASD connection on X with boundary values sufficiently close to Γ has δ-exponential decay to a flat connection. By the compactness of the character variety we can fix $\delta > 0$ such that for any metric $g \in B$ any connection $[\omega]$ in $\mathcal{M}_\ell(X, g)$ there is a product structure for the bundle over the end of X and a flat connection Γ on $N \times SU(2)$ such that in this product structure ω has δ-exponential decay to the pullback of Γ.

According to Proposition 11.0.4 the map $\mathcal{B}^0_{\ell,\delta}(X) \to \mathcal{R}(N)$ constructed in Chapter 7 is a morphism of smoothly stratified spaces which pulls back the local cone bundle neighborhoods. For any metric $g \in B$ the moduli space $\mathcal{M}_\ell^0(X,g)$ is a subset of $\mathcal{B}_{\ell,\delta}^0(X)$. The restriction to $\mathcal{M}_\ell^0(X,g)$ of the natural projection to $\mathcal{R}(N)$ is the boundary map ∂^0.

We can also consider the parameterized moduli space

$$\tilde{\mathcal{M}}_\ell^0(X,B) \subset B \times \mathcal{B}_{\ell,\delta}^0(X).$$

Of course, by Proposition 11.0.3 this space is a smoothly stratified space with local cone bundle neighborhoods pulled back via the the natural projection map

$$\pi\colon B \times \mathcal{B}_{\ell,\delta}^0(X) \to \mathcal{R}(N)$$

from those of $\mathcal{R}(N)$.

CLAIM 14.0.4. *The map*

$$P_+F\colon B \times \mathcal{B}_{\ell,\delta}^0(X) \to \Omega_{1,\delta}^2(X, adE)$$

is a morphism of smoothly stratified spaces. Furthermore, if $([\omega],g) \in \mathcal{M}_\ell^0(X,B)$ *then the restriction to the* $Ker d\pi$ *of the differential of* P_+F *at* $([\omega],g)$ *is surjective.*

Proof. The smoothness is clear. The surjectivity is proved by the usual unique continuation argument as in [12]. □

As a consequence, we see that the parameterized moduli space $\mathcal{M}_\ell^0(X,B)$ is a smoothly stratified space with local cone bundle neighborhoods. The map $\partial^0\colon \mathcal{M}_\ell^0(X,B) \to \mathcal{R}(N)$ is a map of smoothly stratified spaces and pulls back the cone bundles of $\mathcal{R}(N)$ to cone bundle neighborhoods for $\mathcal{M}_\ell^0(X,B)$.

By Sard's theorem, for a generic metric $g \in B$ the moduli space $\mathcal{M}_\ell^0(X,g)$ is a smoothly stratified space with local cone bundle neighborhoods, the boundary map $\partial^0\colon \mathcal{M}_\ell^0(X,g) \to \mathcal{R}(N)$ is a map of smoothly stratified spaces which pulls back the local cone bundle neighborhoods.

The detailed results on local structures of the various $\mathcal{M}^0_\ell(X, g, R_m)$ follow from Theorem 13.1.1. The various dimensions follow directly from Theorem 9.0.1 and the computations in Lemma 13.4.1. □

REMARK 14.0.5. *In the case when $k < 0$ for a generic metric g the based moduli space $\mathcal{M}^0_\ell(X, g)$ locally sits as a C^2 Whitney stratified subset of the thickened moduli space. The reason is that $\mathcal{R}(N)$ locally is a C^2 Whitney stratified subset of $\mathcal{H}_\Gamma \times_{Stab(\Gamma)} \eta_n$ and an open subset of $\mathcal{M}^0_\ell(X, g)$ is identified with the preimage under p^0 of $\mathcal{R}(N)$. By Theorem* 9.0.1 *for a generic metric g the map p^0 will be transverse to $\mathcal{R}(N)$ and hence its preimage will be a C^2 Whitney stratified subset of the based thickened moduli space. It is easy to see that the local C^2 stratifications with local cone bundle neighborhoods induced on $\mathcal{M}^0_\ell(X, g)$ by these local embeddings is compatible with the global smooth stratification with local cone bundle neighborhoods given in the above theorem.*

CHAPTER 15

The case of $SO(3)$ moduli spaces

In this brief chapter we discuss a setting for $SO(3)$-bundles to which the $SU(2)$-theorem of the previous chapter easily generalizes.

First, let us discuss how we define our $SO(3)$-invariants. Here we are following [22]. Let M be a 4-manifold and suppose that $w_2 \in H^2(M; Z/2Z)$ has the property that no $SO(3)$-bundle with this as second Stiefel-Whitney class has a flat connection. We call such classes *essential*. Fix a bundle E with $w_2(E)$ essential and with $p_1(E) = -4\ell$. Fix a generic metric on M. We consider the space of ASD connections on E modulo the group of bundle automorphisms which, when viewed as equivariant maps $E \to SO(3)$, lift to maps $E \to SU(2)$. Equivalently, we could require that the mapping torus bundle over $M \times S^1$ of the gauge transformation has w_2 which annihilates all classes of the form $C \times S^1$ for $C \subset N$. This is a union of components of the full $SO(3)$-gauge group. The moduli space obtained by dividing out by this gauge group is smooth and orientable. We form the compactification of this moduli space by taking Uhlenbeck limits as usual. The fact that w_2 is preserved under these limits together with the fact that $w_2(E)$ is essential means that no Uhlenbeck limit is a flat connection. Thus, for a generic metric the compactified moduli space will be a smoothly stratified space of the correct dimension. We define the Donaldson polynomials $\gamma_{k,w}(M)$ by using the fundamental class of this space and Donaldson's μ-map. We shall not enter here into the issue of what choices must be made to orient this moduli space and hence to determine the sign of $\gamma_{k,w}(M)$, since it is not relevant for our purposes.

Consider a cylindrical manifold (X, g) with end modeled on $[0, \infty) \times N$. Fix an $SO(3)$-bundle E over X again with the property that $w = w_2(E)$ is essential. Suppose that E restricted to a neighborhood of infinity in X lifts to a trivial $SU(2)$-bundle. With the choice of this lift there is a well-defined boundary map $\partial^0 \colon \mathcal{M}^0_{\ell,w}(X, g) \to \mathcal{R}(N)$ where $\mathcal{R}(N)$ is the representation variety of $\pi_1(N)$ into $SU(2)$. In this context the generic metrics theorem Theorem 9.0.1 and the structure theorem Theorem 14.0.1 hold.

CHAPTER 16

A Vanishing Result for 4-manifolds split by certain 3-dimensional nilmanifolds

By a nilmanifold we mean a 3-manifold which is a principal S^1-bundle over the two-torus T^2 with non-trivial Euler class.

Recall that for compact oriented 4-manifold A with boundary we denote by $b_2^+(A)$ the dimension of any maximal positive definite subspace of $H^2(A, \partial A; R)$ for the intersection pairing.

For a closed 4-manifold M we denote by $d(k) = d_M(k)$ the number

$$4k - \frac{3}{4}(\chi(M) + \sigma(M)).$$

It is one-half of the formal dimension of the moduli space of ASD connections on a principal $SU(2)$-bundle over M with Chern class $c_2 = k$.

Here is our main result.

THEOREM 16.0.1. *Let M be a closed oriented 4-manifold and suppose that we have a decomposition $M = P \cup_{\overline{N}} Q$ where $\overline{N}$ is a nilmanifold. Suppose that the Euler class of $\overline{N}$ is even (and nonzero) and that $b_2^+(P), b_2^+(Q) > 0$. Then for all $k > 0$ the Donaldson polynomial invariant $\gamma_k(M)$ associated to $SU(2)$-bundles with $c_2 = k$ vanishes when evaluated on $d(k)$ classes in the image of $H_2(M - \overline{N}; Q) \to H_2(M; Q)$.*

COROLLARY 16.0.2. *Let M be a closed oriented 4-manifold with $b_2^+(M) > 1$*

and $H_1(M) = 0$. Suppose that $T^2 \subset M$ is a smoothly embedded torus with self-intersection ≥ 2. Then for all $k \geq 0$ the Donaldson polynomial invarinat $\gamma_k(M)$ vanishes.

Proof that Theorem 16.0.1 implies Corollary 16.0.2. It follows from [9] that if the Donaldson polynomial invariants for the blowup $\hat{M}$ of M vanish then so do those of M. Clearly, if M has an embedded torus of self-intersection c then $\hat{M}$ has an embedded torus of self-intersection $c-1$. In this way we can reduce to the case when the self-intersection of the torus is positive and even. In this case the boundary of a tubular neighborhood of the torus is a nilmanifold $\overline{N}$ with even Euler class. Since $b_2^+(M) > 1$ and since the self-intersection of the torus is positive, both sides of this nilmanifold in M have positive b_2^+. Furthermore, it is easy to see that the map $H_2(M - \overline{N}; Q) \to H_2(M; Q)$ is onto. The result is now immediate from the theorem.

REMARK 16.0.3. *By the trick of blowing up we were able to change the self-intersection of the torus from odd to even. Since the odd nilmanifolds have irreducible representations into $SU(2)$ we cannot prove directly a vanishing theorem for them. We know of no trick to reduce the odd case to the even case in general, and indeed, we think it is an interesting question whether this result holds for odd nilmanifolds.*

Proof. After choosing an orientation on T^2 (and hence $\overline{N}$) we can evaluate the Euler class on the fundamental class and thus think of it as an integer. We choose an orientation for T^2 so that the Euler class is negative. Let N be the oriented 3-manifold induced from $\overline{N}$ with this orientation. We suppose that the nilmanifold $\overline{N} \subset M$ separates M into two manifolds P and Q. We label the components so that $\partial Q = N$ and $\partial P = -N$.

The plan of the proof goes as follows. Fix an $SU(2)$-bundle $E \to M$ with $c_2(E) = k$. For technical reasons having to do with flat connections, cf, [22], it is simpler to blow up and consider $SO(3)$-invariants. More precisely, consider M', the manifold obtained by forming the connected sum each of P and Q with a copy of $\overline{CP}^2$, and E', the $SO(3)$-bundle obtained by taking the connected sum of adE with the $SO(3)$-bundle over each copy of $\overline{CP}^2$ with $p_1 = -1$. Let $e_1, e_2 \in H^2(M'; Z)$ be the exceptional cohomology classes. Notice that E', $E'|P$, and $E'|Q$ all have essential w_2, and that $E'|\overline{N}$ lifts to an $SU(2)$-bundle. By [22] (also R. Brussee, personal communication) we know that the $SU(2)$-Donaldson polynomial for M constructed from E is one-quarter of the coefficient of e_1e_2 in the $SO(3)$-Donaldson polynomial for M' constructed using the bundle E'. Thus, to prove our theorem it suffices to prove that the Donaldson polynomial of M' constructed using E' vanishes on classes in the image of $H_2(M' - \overline{N}) \to H_2(M')$.

For the rest of the proof we work with SO(3)-connections, and replace M by M' and E by E'. We shall fix a sequence of generic metrics on M which have tube lengths going to infinity across $\overline{N}$. We shall fix classes $\alpha_1, \ldots, \alpha_{d(k)}$ in the image of $H_2(M - \overline{N}) \to H_2(M)$. We then represent these classes by appropriately chosen riemann surfaces $\Sigma_1, \ldots, \Sigma_{d(k)}$ in $M - \overline{N}$. Furthermore, we choose generic divisor representatives Δ_i in the spaces of irreducible connections over the Σ_i. Then we can pull these divisors back to the moduli spaces of ASD connections on the $SO(3)$-bundle over M with respect to the seqeuence of metrics. Doing this gives codimension-two submaifolds of these moduli spaces representing the images of $[\Sigma_i]$ under the μ-map. The value of the Donaldson polynomial on the $\alpha_1, \ldots, \alpha_{d(k)}$ is simply the signed intersection number of these submanifolds. We shall show that for all sufficiently long tube lengths that there are no points of intersection, hence showing that the value of the Donaldson polynomial on these classes is zero. This will prove the result.

Now let us begin with the details of the proof. By the above discussion we can assume that $E \to M$ is an $SO(3)$-bundle so that:

- $E|\overline{N}$ is trivial,

- $p_1(E) = -4k$ with $k > 0$,
- $w_2(E) = w$ is essential
- $w|P$ and $w|Q$ are both essential.

We must show that $\gamma_{k,w}(M)$ as defined in the previous chapter evaluates trivially on collections of $d(k)$ classes in the image of $H_2(M - \overline{N}) \to H_2(M)$.

First, we choose $g_{\overline{N}}$ to be a metric on $\overline{N}$ of the form $d\sigma^2 + \theta \otimes \theta$ where $d\sigma^2$ is a flat metric on T^2 and θ is a connection 1-form for the S^1-bundle over T^2 whose curvature is a multiple of the volume form for $d\sigma^2$. Next, we fix a metric g_0 on M which is the product metric $dt^2 + g_{\overline{N}}$ on a collar $[-2,2] \times \overline{N}$ of $\overline{N}$. Then, we fix a sequence of metrics g_n associated with g_0 on $(M, \overline{N})$ which have tube lengths going to infinity.

By the moduli spaces $\mathcal{M}_{\ell,w|X}(X, g_\infty)$ and $\mathcal{M}_{\ell,w}(M, g_n)$ we mean the moduli spaces of ASD connections on $E|X$ and E with repsect to the given metrics of energy $8\pi^2\ell$, modulo gauge equivalence by gauge transformations which lift to $SU(2)$ as described in the previous chapter. We also have $\mathcal{M}_{\ell,w|P}(P, g_\infty)$ and $\mathcal{M}_{\ell,w|Q}(Q, g_\infty)$. By varying g_0 on balls in $M - [-2,2] \times \overline{N}$ by Theorem 14.0.1, Remark 14.0.3, and the usual generic metrics theorem, we can assume that:

Assumption 1: The geometric limit of these riemannian manifolds is a riemannian manifold (X, g_∞) which has a cylindrical end such that for all $\ell > 0$ Theorem 14.0.1 holds for (P, g_∞) and (Q, g_∞) and for each n the moduli space $\mathcal{M}_{\ell,w}(M, g_n)$ is a smooth manifold.

Let us fix a collection $\alpha_1 \cdots, \alpha_{d(k)}$ of classes in the image of $H_2(M - \overline{N}) \to H_2(M)$. Since the Donaldson polynomial is a rational invariant, we can replace each α_i by any non-zero multiple. Thus, there is no loss of generality in assuming that $\langle w, \alpha_i \rangle = 0$ for all i. For each α_i we fix an embedded surface $\Sigma_i \subset M - ([-2,2] \times \overline{N})$. By Lemma 7.5.1 we can assume that:

Assumption 2: For each i, any irreducible connection A which is either a g_∞-ASD connection on X or a g_n-ASD connection on M for some $n \geq 1$, the restriction $A|\Sigma$

is also irreducible.

Because of Assumption 2, for each i the restriction mapping induces maps

$$r_{i,\infty} : \mathcal{M}_{\ell,w|X}(X, g_\infty) \to \mathcal{B}^*(\Sigma_i)$$

and

$$r_{i,n} : \mathcal{M}_{\ell,w}(M, g_n) \to \mathcal{B}^*(\Sigma_i)$$

where by $\mathcal{B}^*(\Sigma)$ we mean the usual $SU(2)$-gauge equivalence classes of irreducible $SU(2)$-connections on the trivial $SU(2)$-bundle over Σ_i. By the discussion in Section 9.4 these maps are smoothly stratified. There are classes $\mu_i \in H^2(\mathcal{B}^*(\Sigma_i); Z)$ whose pullbacks under the $r_{i,n}$ and $r_{i,\infty}$ are the usual classes $\mu([\Sigma_i])$ in these moduli spaces. By Sard's theorem we can choose codimension-two subvarieties $\Delta_i \subset \mathcal{B}^*(\Sigma_i)$ subvarieties with normal bundles dual to the class μ_i in such a way that the following four assumptions hold:

Assumption 3: For each n the map $r_{i,n}$ are transverse to Δ_i, and the resulting smooth divisors $D_{i,n} \subset \mathcal{M}_{\ell,w}(M, g_n)$ are in general position.

Furthermore, by Lemma 7.5.3 we can arrange that the following two assumptions hold for each i and each $\ell > 0$.

Assumption 4: For the component P of X the map

$$\mathcal{M}_{\ell,w|P}(P, g_\infty) \to \mathcal{B}^*(\Sigma_i)$$

is transverse to Δ_i and the preimages $D_{i,P} \subset \mathcal{M}_{\ell,w|P}(P, g_\infty)$ of the Δ_i are smooth divisors meeting transversely.

We say that the $D_{i,P}$ are *divisors on the P-side.*

Assumption 5: For the component Q of X the restriction of the map

$$\mathcal{M}_{\ell,w|Q}(Q, g_\infty) \to \mathcal{B}^*(\Sigma_i)$$

to any stratum of $\mathcal{M}_{\ell,w|Q}(Q, g_\infty)$ is transverse to Δ_i and the preimages of the Δ_i meet transversely in that stratum.

These are the divisors *on the Q-side.*

Lastly, since the maps $\partial\colon \mathcal{M}_{\ell_1,w|P}(P,g_\infty) \to \chi(\overline{N})$ and $\partial\colon \mathcal{M}_{\ell_2,w|Q}(Q,g_\infty) \to \chi(\overline{N})$ are smooth over the pre-image of the smooth points V of $\chi(\overline{N})$, by Sard's theorem we can also assume that:

Assumption 6: For every $\ell_1, \ell_2 > 0$, the divisors D_i in

$$\mathcal{M}_{\ell_1,w|P}(P,g_\infty,V) \times_V \mathcal{M}_{\ell_2,w|Q}(Q,g_\infty,V)$$

induced by the Δ_i meet transversely.

By Assumption 3, for any $n \geq 1$ the value of $\gamma_{k,w}(M)$ on the classes $(\alpha_1,\dots,\alpha_{d(k)})$ is given by the signed intersection of the divisors

$$D_{1,n} \cap \cdots \cap D_{d(k),n} \subset \mathcal{M}_{k,w}(M,g_n).$$

Thus, the following proposition will establish the theorem.

PROPOSITION 16.0.4. *For all n sufficiently large*

$$D_{1,n} \cap \cdots \cap D_{d(k),n} = \emptyset$$

in $\mathcal{M}_{\ell,w}(M,g_n)$.

Proof. The proof is by contradiction. Suppose that for each $n \geq 1$ there is a point

$$[A_n] \in D_{1,n} \cap \cdots \cap D_{d(k),n}.$$

By Theorem 6.3.3 after passing to a subsequence and adding extra base points in the tube $[-2,2]\times\overline{N}$ there is an Uhlenbeck limit of the $\{[A_n]\}$ without loss of energy. The limit is of the form

$$(a_Q, a_1, a_2, \cdots, a_s, a_P)$$

where $a_P = (\omega_P, x_1, \dots, x_{n_P})$ and $a_Q = (\omega_Q, z_1, \dots, z_{n_Q})$ are generalized g_∞-ASD connections over P and Q and the $a_i = (\omega_i, y_i^1, \dots, y_{n_i}^i)$ are generalized ASD connections over $(-\infty,\infty)\times N$ with respect to the product metric $dt^2 + g_N$. Furthermore, since there is no loss of energy in the limit, by Lemma 6.3.1 the limiting genearlized connection has compatible boundary values. We let ℓ_Q, ℓ_i, ℓ_P be the energies of the generalized connections a_Q, a_i, a_P. Let χ_P, χ_Q be the components of $\chi(N)$ containing $\partial([\omega_P]), \partial([\omega_Q])$, and for each i let χ_i be the component containing $\partial_\infty([\omega_i])$. Lastly, letting $\Gamma_0, \dots, \Gamma_s$ be the connections at infinity given by $[\Gamma_0] = \partial([\omega_Q])$, for $i = 1, \dots, s-1$ we have $[\Gamma_i] = \partial_\infty([\omega_i])$ and $[\Gamma_s] = \partial([\omega_P])$.

The first thing to observe is that because $w|P$ and $w|Q$ are essential, it must be the case that $\omega_P, \omega_Q > 0$ are not flat. Also, if one of the $\ell_i = 0$, then we can drop it from consideration and reduce s by one. In this way we also arrange that in addition each of the $\ell_i > 0$.

Of course, since the divisors that we are using to represent the $\mu([\Sigma_i])$ are closed under Uhlenbeck limits we know that the point represented by the generalized connection $[a_P]$ lies in the intersection of the divisors on the P-side, and similarly for Q. This means that for each i for which $\Sigma_i \subset P$ either the background connection ω_P is contained in the divisor $D_{i,P}$ or some point of positive mass for the idealized curvature of a_P lies in Σ_i. Similar statements hold for the i for which $\Sigma_i \subset Q$.

LEMMA 16.0.5. *Let fd_Q be the formal dimension of the fiber of the map*

$$\partial\colon \mathcal{M}_{\ell_Q, w|Q}(Q, g_\infty) \to \chi_Q,$$

and for each i let fd_i be the formal dimension of the fiber of

$$\partial_\infty\colon \mathcal{M}_{\ell_i}((-\infty,\infty) \times N, dt^2 + g_N) \to \chi_i$$

at at any point in the preimage of $(\partial_{-\infty}(\omega_i), \partial_\infty(\omega_i))$. Finally, let td_P be the dimension of $\mathcal{M}_{\ell_P, w|P}(P, g_\infty, \chi_P)$ as given in Theorem 14.0.1. Then

$$fd_Q + \sum_{i=1}^{s} fd_i + td_P \leq dim\mathcal{M}_{k,w}(M, g_k) - (s+1)$$

and

$$\ell_Q + \sum_{i=1}^{s} \ell_i + \ell_P = k.$$

Proof. The second equation is simply the expression that says that there is no loss of energy in the limit. It follows easily from the formulae given in Theorem 14.0.1 and 84 and the fact that the boundary values are compatible that

$$fd_Q + \sum_{i=1}^{n} fd_i + td_P + \sum_{i=0}^{s} dim(Stab)(\Gamma_i) = dim\mathcal{M}_{k,w}(M, g_k).$$

Since each Γ_i is reducible, its stabilizer has dimension at least one. The lemma is now immediate. □

For this formula to be of use we need to know that the formal fiber dimensions fd_i are all at least one. We have the following result of Taubes, see Section 20 of [28].

LEMMA 16.0.6. *For any* $\ell > 0$, *if* $[\omega] \in \mathcal{M}_\ell((-\infty,\infty) \times N, dt^2 + g_N)$, *then the fiber dimension of* ∂_∞ *at* $[\omega]$ *is at least* 1.

REMARK 16.0.7. *In fact, Taubes shows that the fibers of* ∂_∞ *are smooth of the correct dimension. Since the fibers always have a free action of* R^1, *if they are non-empty they have geometric dimension at least one.*

COROLLARY 16.0.8.

$$fd_Q + td_P \leq dim\mathcal{M}_{k,w}(M, g_n) - (2s+1)$$

Proof. To show this we need only show that $fd_i \geq 1$ for every $i, 1 \leq i \leq s$. This follows immediately from the above lemma if the background connection ω_i is not flat. If the background connection ω_i is flat then $\ell_i \geq 1$ and the boundary values of ω_i at $-\infty$ and ∞ are equal. In this case it is easy to see from Equation 84 that the formal fiber dimension fd_i is $8\ell_i - dim(Stab)(\Gamma_i) \geq 5$. □

Now we are ready to complete the proof of the proposition. We divide into two cases: $s > 0$ and $s = 0$.

Case I: $s > 0$. In this case, it follows immediately from Corollary 16.0.8 that

$$fd_Q + td_P < dim\mathcal{M}_{k,w}(M, g_n) - 2 = 2d(k) - 2. \tag{125}$$

First suppose that Γ_0 is *ad*-trivial. Then by Theorem 14.0.1 the preimage $\partial^{-1}([\Gamma_0])$ is a stratum of $\mathcal{M}_{\ell_Q,w|Q}(Q, g_\infty)$ and its dimension is equal to fd_Q. The point $[a_Q]$ is a generalized connection of energy ℓ_Q with the given boundary value which lies in the intersection of all the divisors on the Q-side. We claim that fd_Q is greater than or equal to twice the number of divisors on the Q-side. If a_Q is a connection, then this is immediate from Assumption 5. Suppose now that a_Q has points of idealized curvature. Let $8\pi^2 t_Q$ be the total mass of the idealized curvature of a_Q. Then $[\omega_Q] \in \mathcal{M}_{\ell_Q - t_Q,w|Q}(Q, g_\infty)$ and $[\omega_Q]$ lies in the fiber over $[\Gamma_0]$. Since ω_Q is not flat, this fiber has dimension $fd_Q - 8t_Q$. By Assumption 5, the connection $[\omega_Q]$ can lie in at most $fd_Q/2 - 4t_Q$ of the divisors on the Q-side. On the other hand the points of idealized curvature can lie in at most $2t_Q$ of the surfaces Σ_i. Thus, the total number of the divisors on the Q-side containing the generalized connection $[a_Q]$ is at most $fd_Q/2 - 2t_Q$. Since $[a_Q]$ lies in all the divisors on the Q-side, this implies that there are at most $fd_Q/2$ of them.

Similarly, $[a_P]$ is a generalized connection of energy ℓ_P in the intersection of all the divisors on the P-side. Again since the background connection ω_P is not flat, it follows from Assumption 4 and an argument similar to the one above that td_P is greater than or equal to twice the number of divisors on the P-side. This implies that $fd_Q + td_P \geq 2d(k)$, which contradicts Equation 125.

The other possiblitiy is that Γ_0 is not *ad*-trivial. In this case, according to Theorem14.0.1 $[\Gamma_0]$ is a smooth point of $\chi(N)$ and the dimension of the stratum of $\mathcal{M}_{\ell_Q,w|Q}(Q, g_\infty)$ containing $\partial^{-1}([\Gamma_0])$ is of dimension $fd_Q + 2$. Arguing as above using Assumptions 4 and 5 and the fact that the background connections ω_P and ω_Q are not flat, we see that

$$fd_Q + 2 + fd_P \geq 2d(k).$$

This again contradicts Equation 125.

Case II: $s = 0$. In this case we must argue somewhat differently. The reason is that in this case Theorem 14.0.1 tells us that the sum of the dimension of $\mathcal{M}_{\ell_P,w|P}(P, g_\infty, \chi_P)$ and the dimension of the largest stratum of $\mathcal{M}_{\ell_Q,w|Q}(Q, g_\infty, \chi_Q)$ is one more than the dimension of $\mathcal{M}_{k,w}(M, g_n)$.

Part of the previous argument does still apply in this case. Namely, Corollary 16.0.8 tells us that $fd_Q + td_P < 2d(k)$. Thus, as before we can rule out the possibility that Γ_0 is *ad*-trivial.

We now consider the case when Γ_0 is not *ad*-trivial. Since $s = 0$ and since the limit has compatible boundary conditions, it follows that $\partial([\omega_Q]) = \partial([\omega_P])$. Let t_P and t_Q be $(8\pi^2)^{-1}$ times the total idealized energy in the generalized connections $[a_P]$ and $[a_Q]$. Thus, the point $([\omega_Q], [\omega_P])$ is a point of the fibered product

$$\mathcal{M}_{\ell_Q - t_Q,w|Q}(Q, g_\infty, V) \times_V \mathcal{M}_{\ell_P - t_P,w|P}(P, g_\infty, V)$$

where $V \subset \chi(N)$ is the set of smooth points. Since ω_P and ω_Q are both not flat, it follows from Theorem 14.0.1 that this fibered product is a smooth manifold of dimension $2d(k) - 1 - 8t_P - 8t_Q$. At most $2t_P + 2t_Q$ of the surfaces Σ_i can contain points of idealized curvature of a_P and a_Q. Thus, for $[a_P]$ to lie in the intersection of the divisors on the P-side and for $[a_Q]$ to lie in the intersection of the divisors on the Q-side, it must be the case that there are at least $d(k) - 2t_P - 2t_Q$ of the divisors with non-empty intersection in the fibered product. By Assumption 6, this is impossible since

$$2d(k) - 4t_P - 4t_Q > 2d(k) - 1 - 8t_P - 8t_Q.$$

We have reached a contradition in all possible cases. This contradiction completes the proof of the proposition. $\square$

As we have already observed, the theorem follows immediately from the proposition.

Bibliography

[1] M. Atiyah, N. Hitchin, and I. Singer, *Self-duality in four-dimensional Riemannian geometry*, Proc. R. Soc. London, **362** (1978), 425–461.

[2] M. Atiyah, V. Patodi, and I. Singer, *Spectral asymmetry and Riemannian geometry: I*, Math. Proc. Camb. Phil. Soc., **77** (1975), 43–69.

[3] ———, *Spectral asymmetry and Riemannian geometry: II*, Math. Proc. Camb. Phil. Soc., **78** (1975), 405–432.

[4] ———, *Spectral asymmetry and Riemannian geometry: III*, Math. Proc. Camb. Phil. Soc., **79** (1976), 71–99.

[5] M. Atiyah and I. Singer, *The index of elliptic operators: I*, Annals of Math., **87** (1968), 484–530.

[6] D. Auckly, "Computing secondary and spectral invariants", PhD thesis, University of Michigan, 1991.

[7] A. Casson and C. Gordon, *Cobordism of classical knots*, in "À la recherche de la Topologie Perdue", A. Marin and L. Guillou, eds., Progress in Mathematics, Birkhauser, Boston, 1986.

[8] S. Chern and J. Simons, *Characteristic forms and geometric invariants*, Annals of Math., **99** (1974), 48–69.

[9] S. Donaldson, *Polynomial invariants of smooth 4-manifolds*, Topology, **29** (1990), 257–315.

[10] S. Donaldson and P. Kronheimer, "The Geometry of Four-Manifolds", Clarendon Press, Oxford, 1990.

[11] J.-P. Eckmann and C. Wayne, *Propagating fronts and the center manifold theorem*, Comm. Math. Phys., **136** (1991), 285–307.

[12] D. Freed and K. Uhlenbeck, "Instantons and 4-Manifolds", MSRI Lecture notes, Springer-Verlag, 1984.

[13] W. Goldman and J. Millson, *The deformation theory of representations of fundamental groups of compact Kähler manifolds*, Publ. I.H.E.S., **67** (1988), 43–96.

[14] ———, *The homotopy invariance of the Kuranishi space*, Ill. J. Math., **34** (1990), 337–367.

[15] R. Gompf and T. Mrowka, *Irreducible four manifolds need not be complex*, Annals of Math., **138** (1993), 61–111.

[16] L. Hörmander, "The analysis of linear partial differential operators", Grundlehren der Mathematischen Wissenschaften, Springer–Verlag, New York, 1983–1985.

[17] L. Kauffman, "On Knots", vol. 115 of Annals of Math Studies, Princeton University Press, Princeton, New Jersey, 1987.

[18] P. Kirk and E. Klassen, *Chern-Simons invariants of 3-manifolds and representation spaces of knot groups*, Math. Ann., **287** (1990), 343–367.

[19] P. Kronheimer and T. Mrowka, *Gauge theory for embedded surfaces*, I, Topology, **32**, (1993), 773–826, and II, Topology, to appear

[20] R. Lockhart and R. McOwen, *Elliptic differential operators on non–compact manifolds*, Annali di Scuola Norm. Sup. de Pisa, (1985), 409–448.

[21] G. Matić, *SO(3)-connections and rational homology cobordisms*, J. Diff. Geo., **28** (1988), 277–307.

[22] J. Morgan and T. Mrowka, *A note on Donaldson's polynomial invariants*, Internat. Math. Res. Notices, (1992), 223–230.

[23] T. Mrowka, "A local Mayer–Vietoris principle for Yang–Mills moduli spaces", PhD thesis, U.C. Berkeley, 1989.

[24] D. Ruberman, *Rational homology cobordisms of rational space forms*, Topology, **27** (1988), 401–414.

[25] L. Simon, *Asymptotics for a class of non–linear evolution equations with applications to geometric problems*, Annals of Math., **118** (1983), 525–571.

[26] C. Taubes, *Gauge theory on asymptotically periodic 4-manifolds*, J. Diff. Geo., **25** (1987), 363–430.

[27] ———, *Casson's invariant and gauge theory*, J. Diff. Geo., **31** (1990), 547–599.

[28] ———, *L^2–moduli spaces on open 4–manifolds*, International Press, Cambridge, (1993)

[29] K. Uhlenbeck, *Connections with L^p-bounds on curvature*, Comm. Math. Phys., **83** (1982), 31–42.

Index